AMBER WAVES

AMBER WAVES

THE

Extraordinary Biography OF *Wheat,* FROM *Wild Grass* TO *World Megacrop*

CATHERINE ZABINSKI

THE UNIVERSITY OF CHICAGO PRESS
CHICAGO AND LONDON

Illustrations by Angela Mele

The University of Chicago Press, Chicago 60637
The University of Chicago Press, Ltd., London

Published 2020
Printed in the United States of America

29 28 27 26 25 24 23 22 21 20   1 2 3 4 5

ISBN-13: 978-0-226-55371-9 (cloth)
ISBN-13: 978-0-226-55595-9 (e-book)
DOI: https://doi.org/10.7208/chicago/9780226555959.001.0001

Library of Congress Cataloging-in-Publication Data

Names: Zabinski, Catherine, author.
Title: Amber waves : the extraordinary biography of wheat, from wild grass to world megacrop / Catherine Zabinski.
Description: Chicago : University of Chicago Press, 2020. | Includes bibliographical references and index.
Identifiers: LCCN 2019053515 | ISBN 9780226553719 (cloth) | ISBN 9780226555959 (ebook)
Subjects: LCSH: Wheat—History. | Wheat—Breeding—History.
Classification: LCC SB191.W5 Z27 2020 | DDC 633.1/1—dc23
LC record available at https://lccn.loc.gov/2019053515

∞ This paper meets the requirements of ANSI/NISO Z39.48–1992 (Permanence of Paper).

for Mona *and* Timothy

Contents

AMBER WAVES

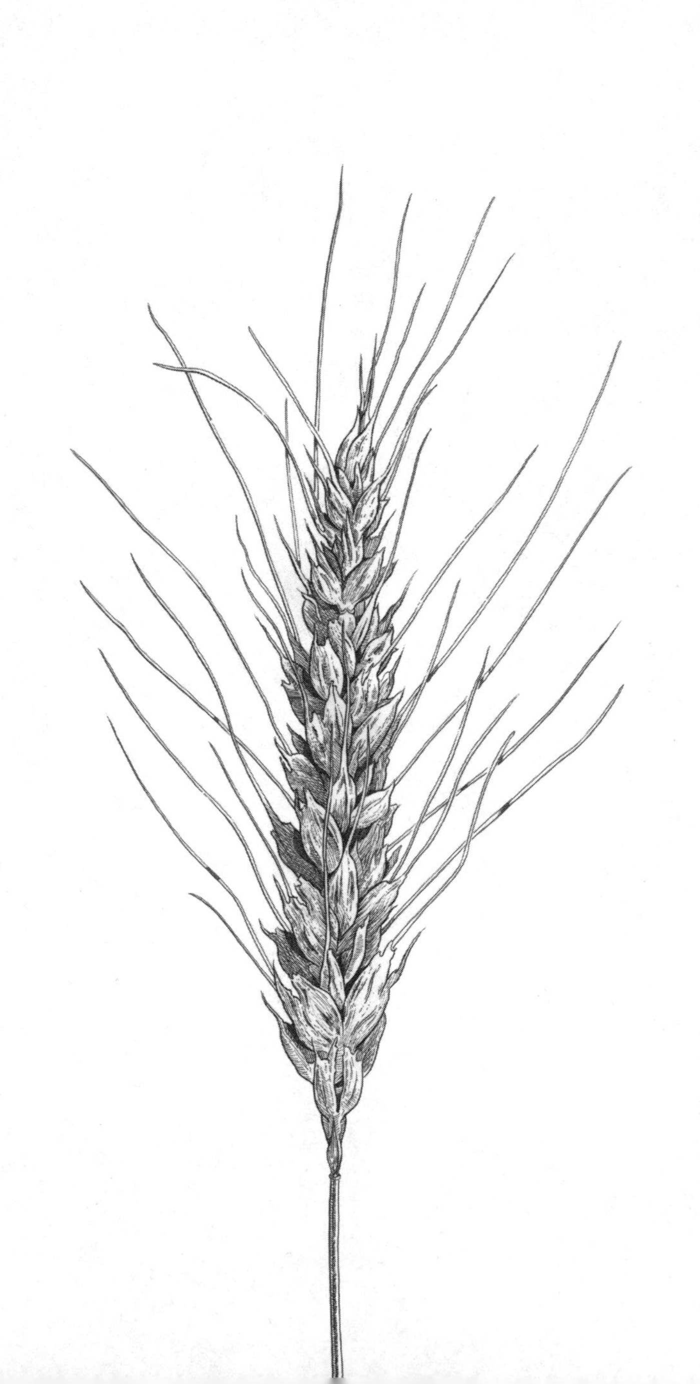

INTRODUCTION

A Biography of Wheat?

A biography of wheat is a bit of a fallacy from the outset, for wheat isn't a person. Moreover, it's not even a single species. Wheat is a group of species, and the one that comes to mind, the plant that we use to make bread and croissants and piecrusts, didn't even exist until several thousand years after our ancestors started cultivating its relatives. And despite their rather bland appearances, the species in this group are quirky. Like the people that move in next door, easily summed up based on outward appearances, the wheats come alive as you get to know them, with all their flukes and foibles.

But still—why read a biography of a group of plants? Ask your stomach that question. Each time you get up out of your chair to forage for something in the kitchen, and every meal you sit down to eat, savoring the aroma before your first bite, are part of the answer. Every child that goes to bed hungry, every family that runs out of money for groceries before the next paycheck, and every pile of food we discard as waste instead of nourishment, are part of the answer. We have learned how to grow food with increasing efficiency (with thousands of years of practice), but at the same time, we organize our societies in complex ways that can make food production and distribution less than straightforward. Getting to know our food species and understanding our relationship with those species are ways to learn more about ourselves.

Because wheat is a group of grasses that our ancestors used

for nourishment, the story of wheat is also a story of human capacity to find a secure and nutritious source of food for energy, for health, for pleasure. A biography of wheat reflects our ingenuity to cultivate and harvest a plant that we then use to make food—our use of wood and rocks and metal, and of human labor, animal labor, and diesel engines. The story of this group of grasses runs parallel to that of our species' early villages, first states, and plans for war and plans for peace. It's also a story of our power and our human proclivity to ensure our security by overpowering others. And that's easily done by controlling people's access to the food that we all need to survive and prosper.

Further, it's a story about science, about soil and roots, leaves and sunshine, and water and photosynthesis. Our ancestors used simple observation and reasoning and trial and error to grow food, and as technical skills increased, chemistry labs and greenhouses to understand how to grow more food more efficiently. Our bodies need a minimum number of calories, a specific set of chemical elements, to maintain and support our muscles, our blood, and the rest of our anatomy. Our brains let us know if we don't have enough of what we need. With taste and smell, we discriminate between good and bad food to keep us from eating something that will make us sick or worse. As much as we avoid food that we think is no good, we are drawn to food that we savor. For many, the smell of freshly baked bread or the sight of a plate of steaming pasta is enough to trigger a mouthwatering response and the anticipation of a moment of bliss. But nothing about food is simple, because humans aren't simple. For some of us, a warm cinnamon roll does not inspire ecstasy; wheat triggers a physical response in the gut that makes simple repeated exposure a potentially lethal prospect.

As a species, we invest energy to secure food for ourselves and for our young. As our societies became more structured and our agriculture more technology driven, fewer of us are actually involved in growing the plants and raising the animals that we eat. But still, we shop, we cook, we order online, we earn money to pay for the food we eat. A biography of wheat, then, is a story of a group of grasses whose existence became complicated by its convergence with our own species and our never-ending need for more food.

The other problem with a biography of wheat is that it inevitably distorts time. Wheat and its relatives had lived for millions of years on this planet before our ancestors were alive to pay any attention to them. So to put the parts of the story in which we're most interested—those parts where we're also in the picture—into a context, the biography of wheat covers a really long period. The first chapter of this story spreads over a couple of billion years, the second chapter over millions of years, and the third over thousands. A tale with an evolutionary backdrop requires some flexibility in thinking about time. And that evolutionary backdrop is essential to the story. It's essential because of our concerns about the safety and security of our food. It's not just a personal task to keep our cupboards stocked and some money in our checking account for groceries; it's also a political task to provide enough food so that people are not rebelling because of hunger. And the technological tools that are increasingly applied, not just to food production but in our hospitals and research labs, are focused on the molecular level. Our capacity to add or edit genes, not only in our own DNA but in the DNA of our food species, is best understood in the context of the mechanisms of evolution. Accordingly, this biography will take you from the origins of the first plants up to the current days; it's about the evolution of plants, the ways our ancestors used the tiny grass seeds to feed themselves, and the challenges of agriculture in a world where our population doubles in less than a century. This biography travels the globe, because our species has transported wheat to all but the polar continents. A biography of wheat, then, is not just about the plants. It's also about us, our societies, and how we manage our food.

CHAPTER ONE

The Whispering of the Grasses

In the lengthening shadows of the early evening, the warm breeze traces the contours of the foothills. It's quiet here; the whispering of the grasses and the tapping of the oak leaves and an occasional twittering of a lone nuthatch caching seeds are the only sounds to break the silence. The grasses growing amid the scattered oak trees are like grasses you find anywhere: long narrow leaves, humble flowers—no petals, just the business end exposed to the wind—and roots growing through the rocky soils to forage for whatever food and water this site has to offer. But these grasses aren't like any other. They belong to species that will change the world. And the rolling foothills at the base of the Zagros Mountains are in a part of the world called the Fertile Crescent, home to one of the sites where agriculture originated.

Grasses don't have brains. They don't think, they don't plan, they don't scheme. If they did, the grass plants under the oak trees on this warm evening might be credited with planning their empire, conjuring strategies to harness the energy of the most powerful species on earth. The grasses growing on these quiet hills in Turkey are ancestors of the wheat that grows in the lowlands of the Rocky Mountains, the humid fields of Kansas, the grain farms of northern France, and the northern plains of India. If these grasses had brains, they hardly could have planned a better strategy to ingratiate themselves to humans. The descendants of these grasses will be transported around the globe, carefully planted, watered, and tended for generations to come. They will be honored with Greek and Roman deities, enshrined by Old World painters, woven into national ballads, and stamped onto coins. The descendants of these grasses will become commodities for people worldwide; they will be bought and sold, traded and exchanged, worried and fought over. These grasses will be used as a weapon during wartime, stolen from neighboring countries and withheld from enemies. Equally, they will be grown and dispersed in an effort to maintain peace in poor countries. They will be venerated and they will be vilified for the properties of their seeds. If the grasses on this hill had brains, the whistling wind might have heard a voice of dissent on the edge of the field, saying, "Be careful who your friends are. We have it good now. Hang on to the simple life."

In fact, the evolutionary pathway of these grasses from this point onward will be largely directed by humans. The people who gathered seeds in the oak savannas had learned that they could grind the seeds with a rock, make that powder into a paste, and be nourished. And once the power of the grasses to nurture was understood, our ancestors carried grass seeds with them as they moved from place to place. Planted in new soils, with a new climate, the grasses either adapted or they died. Without a knowledge of evolutionary mechanisms, of selection pressures, or of DNA, humans carried the grass seeds in all directions from the hillslopes of Turkey. And in those new environments, the grasses differentiated into varieties that grow well in the rocky soils of the Balkan Peninsula, in the cool springs of Scandinavia, and in the humid summers of India. Partly because of the genetic flexibility of grasses, and partly because humans have always been self-interested agents of change for their food species, the future for those hillslope grasses was launched on an improbable trajectory.

The story of wheat is intricately linked with our own, at least the last 10,000 years of that account. The parts of a story that include us are always the most interesting, but in fairness to wheat, the subject of this biography, we'll start at the beginning. The story of wheat begins with its ancestors, those species that developed and passed on the traits that enabled these grasses to attract our attention so that we would spread them across the globe. We go back in time to when the only plants were tiny, waterborne things: single-celled green algae, the kind you can find growing as a layer of scum in a quiet area of a pond. These green algae are simultaneously unremarkable and miraculous—unremarkable for their size and inconspicuousness, yet miraculous for their capacity to photosynthesize. Photosynthesis is the plant's version of Midas's golden touch: sunlight is converted into chemical bond energy, the currency for life.

That conversion happens like this: a ray of sunlight hits a green chlorophyll molecule embedded in an alga. But that light energy is a little too much for the chlorophyll, and in the process of absorbing the energy, one of its electrons, the negatively charged particles that orbit atoms, jumps out of its orbit. What ensues looks like a game of hot potato, with a negatively charged electron bouncing from one molecule to another, till something (a molecule called NADPH [nicotinamide adenine dinucleotide phosphate]) finally catches it and holds on to it. Meanwhile, the chlorophyll molecule at the start of this chain is short one electron, so it pulls one from a water molecule, causing the water to split down the middle, releasing oxygen into the atmosphere.

The hot-potato movement of negatively charged electrons causes an electric charge accumulation, similar to when you drag your stockinged feet across the carpet. But instead of generating sparks, the energy from that charge buildup gets transferred to another storage molecule, this one an ATP (adenosine triphosphate), which functions like the emergency granola bar you keep in the bottom of your backpack. The algal cell stores the energy till it's needed, and in the story of photosynthesis, what it's needed for is to use atmospheric carbon dioxide (CO_2) to build sugars and starches. Those sugars and starches are what make up the bulk of plants, the stems, leaves, and roots of the forests and grasslands and aquatic plant life. That's a big thing, in and of itself, but photosynthesis also altered the chemistry of the planet. The oxygen, released in the first half of photosynthesis, accumulated in the atmosphere, a change that altered the possibilities for evolution of future life-forms.[1] We aren't the only ones with an outsize effect on our planet.

Photosynthesis may be the hallmark of the plant kingdom, but it was in fact filched from another group: the cyanobacteria, more affectionately known as blue-green algae. Bacteria can live just about anywhere if a little moisture and some source of food are available. Because of their abundance, bacteria make for a good and reliable food source for other single-celled organisms, such as protists. Protists feed by wrapping their blobby mass around their prey, often bacteria, first engulfing the microbe and then excreting enzymes that dissolve it into its nutrient components. But one time, about a billion and a half years ago, before dissolving the bacteria into oblivion, the protist paused and let the bacteria be. Maybe it was just stocking up, like a big grocery trip on the weekend—no need to eat all the food right away. The bacteria remained alive within the protist's body, even reproducing within its host. When the protist divided to make two daughter cells, copies of the bacteria got passed on to each of them.

For the bacteria trapped inside protists, why not break out, and reclaim their independence? We don't really know why that didn't happen a billion and a half years ago. But one thing that did happen is something we have spent a lot of money and many years working to perfect for our own purposes—DNA transfer from one organism to another. Bacterial DNA moved into the nucleus of the protist, so that genes necessary for the bacteria to reproduce became part of the protist's DNA. That transfer eliminated the possibility for independent living by the bacteria, and was an important step in the transition from being a free agent (albeit a bacterium) to being a chloroplast inside another cell. The takeover of photosynthetic bacteria, relegating them to function as chloroplasts, has happened more than once throughout time, spawning different branches of the algal family; but the blue-green algae are the ones that gave rise to the rest of the plant kingdom.

The next step on the evolutionary road to wheat is the development of plants that are made up of more than just a single cell. This is big, because the success or failure of the new organism, multicellular algae, is determined by the capacity for communication and cooperation between its cells. For that, the cells adhere to one another thanks to a sugary gel that surrounds them. Tiny straw-like tubes penetrate the gel, linking each cell to its neighbors. Nutrients move between cells, and signal molecules let one cell know what the other is doing. Cells in a multicellular organism differentiate, a way to split the workload. For example, algal cells on the outside of a multicellular Volvox have flagella, little whiplike structures that with some coordination can propel algae through the water, while inside are large cells devoted entirely to reproduction. Both cell types in the Volvox contain the same DNA, and the mechanics of how the DNA stores and uses information is consistent between cells: DNA gets transcribed to a messenger RNA particle, which gets translated into a series of amino acids which make up a protein. But the DNA is regulated so that certain of its sections are transcribed and others ignored, resulting in cells with different functions. Our eye cells and those of our skin contain the same DNA, but different genes are expressed in each cell type. What changed between the single-celled and the multicellular algae is neither the amount of DNA nor a difference in capacity to generate proteins, but rather how the DNA is regulated. Some genes are left on, meaning they could get transcribed and translated into proteins, and others are turned off, blocked somehow so the cellular machinery ignores them. Gene regulation plays a major role in evolutionary changes—this hypothesis was proposed over 40 years ago—but many of the intricacies of gene regulation still elude us.[2]

Although they both photosynthesize and are multicellular, algae and wheat are still worlds apart. Wheat, for one, grows in soil; it doesn't float in ponds. The transition to land was not an easy one for wheat's ancestors. Water makes for a good home. With its buoyancy, it provides physical support, so organisms don't need structural tissues, like wood or bones. Because nutrients dissolve in water, algae swim in their food source, making nutrient acquisition straightforward. And the water protected algae from UV radiation—present at high levels in the thin atmosphere of 500 million years ago—and buffered daily temperature fluctuations. The move to land required early plants to manage drying-down periods, high UV levels, stronger gravitational forces, and the need to forage for nutrients. Yet despite all those disincentives, colonizing land environments would mean no competition from other plants. CO_2 is easier for plants to access from the atmosphere than from water, and sunlight is more direct and less attenuated on land, so there's more energy to drive photosynthesis. The first terrestrial plants likely found themselves on land after the pool they were floating in dried up for part of the year. Plants that could figure something out for the drying-down periods had a definite competitive advantage.

The shift from buoyant water to dry land was made possible by a trait shared by every member of the plant kingdom: the cell wall. The contents of all our cells are held inside a membrane, like the plastic bag used to carry newly purchased goldfish home to the aquarium. If the cell membrane is the plastic bag that holds water and goldfish, then the plant cell wall is a shoe box used to carry the plastic bag. It is a peculiarity for the plants, some bacteria, and fungi; our own cells have membranes but no walls. The cell wall is in fact why we can make paper from trees and what makes plants good candidates for biofuels.

The cell wall is a highly organized structure (the shoe box analogy fails here by giving it an unappealing simplicity); up to 10% of a plant's genes are devoted to coding for proteins for the composition and metabolism of the cell wall. It's composed of a mixture of gels and crystalline structures and bridges. Molecules of glucose, one of the sugars made during photosynthesis, get linked, one after another, onto a chain that's maybe a couple thousand glucose molecules long. These strands, called cellulose, line up side by side, and the hydrogen atoms in the cellulose link arms, so to speak, producing microscopic fibers. Cellulose fibers, when arranged properly, can take the form of a cotton T-shirt or a sheet of paper. In the cell wall, those fibers are arranged with an obsessive-compulsive attention to order, one layer after another, with each layer aligned at a different angle from the previous layer, and suspended in a carbohydrate gel containing some long proteins for bridging the layers.

The layers of the cellulose fibers give our shoe-box wall structural rigidity, compensating for the loss of buoyancy when plants moved to land. But these layers still allow nutrients and water to reach the cell membrane inside. Proteins embedded in the lattice framework perform many functions: some help with communication between cells; others function like bouncers, deciding who warrants a pass into the cell and who gets blocked; and some protect the cell contents from UV radiation, also key to survival in those early environments. If photosynthesis marked the beginning of a major shift in the composition of our atmosphere, colonization of the land was the start of a major shift in the composition of the earth's crust. From a rocky surface to the first layers of soil, the earliest ancestors of wheat helped engineer our planet into one suitable for habitation.

After the algae, the earliest land plants were liverworts and hornworts and mosses. They grow low to the ground, a layer of tissue a couple of cells thick, without roots or stems or leaves. Scientists describe this group based on what it lacks rather than the more esteem-building alternative. These nonvascular plants don't have the plumbing tissue (vascular system) that enables water to reach the top of a redwood tree, or photosynthesis-derived sugars to move from the leaves to the depths of the wheat roots. But even so, these early plants had a significant impact on the earth's atmosphere. Around 444 million years ago, atmospheric CO_2 levels plummeted, contributing to the onset of glaciation. Scientists have proposed that it was the mosses and the lichens that triggered the atmospheric changes. These diminutive plants straddled the rocky surfaces. Without roots, which would have been fairly pointless in the rocky crust, the nonvascular plants excreted acids that dissolved the rocks' surface layer, releasing nutrients. Some of the nutrients were absorbed by the plants, while others were carried by rain and rivers to the ocean, increasing the food source for algae. The algae then absorbed more atmospheric CO_2. In addition to plant uptake of CO_2, the weathering of the rocks by nonvascular plants exposed compounds that react with the atmosphere, binding CO_2. These hypotheses will likely be debated and revised, but they recognize the multiple ways that early plants could have had an outsize effect.[3]

The development of vascular tissues that gave life on land some depth and breadth happened over the next 50 million years. The cell walls, while beautifully engineered to be both strong and porous, were now reinforced with a secondary cell wall that got laid down in helical loops. This wall, like the primary wall on steroids, starts with the same cellulose microfibrils. But instead of being laid down in a matrix of carbohydrates, it's laid down in a matrix of lignin.

Lignin is a big, messy molecule, sprawling out in every direction at once like a midwestern city without a growth plan. Lignin starts out simply with one of the amino acids (phenylalanine), something that plants but not animals manufacture regularly. Then the cell tacks on a hexagonal carbon compound and a short linear carbon compound to the phenylalanine, resulting in a compound (technical name phenylpropanoid) that plants use for all sorts of things, including protection from UV light. Phenylpropanoids then run through different biochemical pathways to generate even bigger and longer compounds. The cell pools the large compounds from different pathways, throws in a few miscellaneous products, like making soup with whatever is in your refrigerator, and creates lignin. Lignin composition varies between one plant species and the next, between cell types within a plant, and even within a single cell. But in all cases, it's a formidable matrix for a cell wall, into which the cellulose microfibrils get laid down.

Why lignin matters is clear when you think about moving water through plants. The tubes for moving water from the roots to the leaves are made of xylem cells. Xylem is made of hollowed-out cells that line up, one on top of another, to create a tube that extends from the deepest roots to the highest leaf. Substituting lignin for the traditional carbohydrate matrix adds strength to the xylem, which is crucial for the walls of the tubes that move water. Consider for example a straw wrapped in paper. You could in fact use the round paper cover like a straw, simply by trimming the ends of the sealed wrapper. But if you tried to suck air through that tube, the paper would collapse on itself, blocking the movement of air. Lignin added to cell walls meant that water could be drawn up through the tubes without the tubes collapsing. The reinforced wall structure was now strong enough to withstand the negative pressure of water moving up through the tubes. Not to mention that those tubes are structural support for the plant body so that plants can grow upward from the ground. With vascular tissue, plants have both the physical support and the infrastructure to grow tall, increasing their capacity to engineer the earth's surface.

The vascular system first appeared in the lycopods, plants that were almost all stem with just tiny leaves. Most of the lycopods have gone extinct; but in their heyday 300 million years ago, in a world before reptiles and mammals appeared, they grew up to 100 feet tall. They grew in steamy, swampy forests, with 6-foot centipedes scuttling across dry surfaces and giant dragonflies buzzing between their trunks. Those forests contributed to the coal deposits that fueled our Industrial Revolution and that we continue to argue over, debating whether it's more important to provide jobs to coal-mining communities or shift to an energy source that doesn't release CO_2 into the atmosphere. The same molecules that enabled the lycopods to grow tall, the lignin and the cellulose of the cell walls, were slow to break down, particularly in waterlogged soils. Over time, as sediments accumulated on top of the undecomposed plant material, the heat and pressure converted lycopod cell walls into coal deposits.

Growing at the same time in the steamy forest swamps were the first plants with true leaves, the ferns. Some ferns were (and still are) the size of trees, and their trunks, instead of being woody like that of a tree, were a rather messy structure—a bunch of stems held together with roots around the outside perimeter. Ferns are the first plants with all the parts that make them recognizable to us: they're green, you can see them from a distance, and they have stems, leaves, and roots.

Yet the algae, the mosses, the lycopods, and the ferns all lack the single thing that inspired the writing of this book, the single reason why wheat and other grasses have been so important to us. None of these plants have seeds. They reproduce via spores. The fern spore, like our human sperm and egg cells, has only half the number of chromosomes as the cells in the fern body. Ferns release millions of spores into the world, ensuring that at least some will land in a moist and protected area. When that happens, the spore cells divide and multiply to produce a platform—like an oil-drilling platform in the waters off the coast—except the platform is extremely small and green. On the bottom of the platform (or for some species on the top), two structures grow: the plant equivalent of a scrotum, and a cup that holds the ovule, the plant egg. If all goes well, a sperm equivalent fertilizes the ovule, and a tiny fern, containing two copies of each chromosome, grows from that platform into a full-size plant.

Compared with reproduction by spores, seeds are a really big deal. First of all, instead of spore cells generating an independent living platform for the development of the sex cells, in the early seed plants that development happened within a new structure: a cone. Male cones produce pollen, and the female cones house the ovules, which develop into a seed when fertilized. Besides the developing embryo, the seed manufactures a reserve made up of starches and proteins that supply energy to the embryo during germination. It's like a spore with a whole support crew, but also with chromosomes from both parents. At maturity and before leaving the parent plant, seeds lose water. Many go into a dormant mode, giving them time to be dispersed away from the parent plant by wind or water or animals, awaiting environmental cues that the conditions are right for germination. The seed has everything the embryo needs to grow, except for the water it imbibes as the first step of germination. The nutrient-rich package that supports plant embryos so well is also a good nutrient source for animals, hence our interest in wheat.

The early seed plants are ones most of us recognize—the cone-bearing pines, firs, spruces, and cedars, known as conifers—along with a group with a lot of personality but less well known, the cycads. The cycads can resemble a palm tree—a trunk with big fernlike leaves at the top. Or sometimes the trunk is compressed, and it's just the leaves you see. The cycads and the conifers had one other thing besides seeds that has been fundamental for us. The gymnosperms, as this group is called, have a way to add more vascular tissue to their stems, one layer after another, to increase the girth. Wood is vascular tissue and, with all that lignin laid down around each cell, is structurally very solid. The tallest tree is 371 feet taller than the tallest human. We rely on a skeletal system for support and a muscular system to move us, but the limitations of a flexible support system mean that we are tiny compared with the towering trees.

The last major structure for the ancestors of wheat to develop and pass on was one that normally doesn't come to mind when you think of the grass family: the flower. About 200 million years passed between the evolution of cones and the evolution of flowers. What distinguishes flowers from cones is that flowers protect the female reproductive tissue, the ovule, inside their sepals. Sepals are similar to petals, but if you think of a red-petaled rose, the sepals are the green outer covering of the opening flower. Flowers help plants attract birds and insects such as butterflies to move their pollen from one plant to another, and in that way help with the whole process of seed production. Consequently, flowers have adapted an array of colors, shapes, landing platforms, and rewards, including a store of sweet nectar, to manage (or manipulate) the insects and birds and small mammals that serve as the go-between for the male and female sex cells.

But when you close your eyes and think about flowers, what do you imagine? Perhaps a water lily, with its large and simple flower, a good example of one of the earliest flowering plants. Or maybe a rose or a tulip, among the Platonic ideals of a flower. Chances are very high, though, that you didn't think of a grass flower. Grasses do have flowers, but they are small and green. Grasses don't need showy flowers to attract bees or birds, because they rely on wind to move their pollen from the male organ to a receptive female surface. Although their reproductive parts are arranged a little differently from those of water lilies, roses, and tulips, the grasses, including wheat, have flowers and belong to this most advanced group of plants.

The ancestors of wheat, passing on genetic gifts of chloroplasts, cell walls, vascular tissue, seeds, and flowers, spent incomprehensibly long periods of time developing this inheritance. Plants moved onto land 500 million years ago, and the first plants containing vascular tissue appeared 50 million years after that. Seeds arose 400 million years ago, flowers 250 million years ago, and the first grasses, which didn't really look much like wheat, appeared about 40 million years ago. Those oh-so-gradual changes were driven by two things: variation and selection. Many evolutionary stories, whether it's the long, long history of wheat's ancestors or our efforts to grow more wheat for our ever-burgeoning population, are driven by variation and selection. Selection can be caused by the environment—which algae can survive when the water in the pool dries up—or it can be caused by us—which of the wheat plants in the field produces the most seed. But selection can work only if there is variation within the group.

Variation is the result of many different mechanisms that operate at the molecular, the cellular, or a larger synthetic level. It begins with the same mechanisms that enable traits to get passed on from one generation to the next: genes. Genes are found on chromosomes; chromosomes are made of DNA, the foundation of heredity. An example of simple elegance, DNA is structured like a twisting ladder, with each rung of the ladder connecting two of four possible nucleotides (the basic structural units of nucleic acids, such as DNA and RNA): adenine, cytosine, guanine, or thymine, known in shorthand as A, C, G, and T. The structure of the ladder results from some basic pairing rules—A always pairs with T to form a rung, and C with G. The information that codes for a gene comes from the order of nucleotides on just one side of the ladder.

The transcription of the nucleotides into RNA, a molecule

that then gets translated into amino acids to make proteins, is the primary mechanism for passing down the information for how a plant or animal or bacterial body is formed. For this system to work well, information must be duplicated accurately, with a system in place for correcting errors. Our cells and the cells of even the simplest organisms do that. But cells also make copying errors that don't get fixed. Sometimes they make two copies of the same section of DNA. The cellular equivalent of looking up from your book while reading, losing your place, and then rereading a sentence or two can result in the mistake of an extra copy of the same gene. The second copy is extraneous, and thus ignored by the editing machinery that fixes any deviations. Over time, however, some of the gene copies change enough to code for a new protein, a protein that interacts with the standard biochemical pathways in a slightly different way. That's just one way, of many different mechanisms, to generate variation.

The evolution of plants, from single-celled algae to the forests and fields of trees and grasses, is partly the result of evolutionary changes in biochemical pathways. A possible scenario: a simple amino acid is channeled down a new biochemical pathway that adds a few carbons to the protein, and the new compound is able to deflect UV radiation. Cells with that new compound are more successful than the others. With another change to the new compound, maybe 100,000 years later, you now have something that's toxic to the cell. Any plant with that change dies. But in one of the plants, the toxic compound combines immediately with another, and it's no longer toxic. And 20 million years later, that same chemical will channel down another biochemical pathway, and so on, until it is modified such that when used as the matrix for cell walls, it changes how plants grow. The evolution of lignin, which meant that trees could grow tall, came about because the cellular machinery ensures not only the production of faithful copies but also the generation of new variation that is the lifeblood of evolution.

The DNA sequence and its potential for mutations, the regulation of each of the genes (whether they are turned on or off), the numbers of copies of genes and whether their DNA sequence is checked for errors and corrected during chromosome duplication—these are just a few of the mechanisms for generating variation. That variation has the potential to generate enough changes that we can trace a journey, albeit fuzzy and incomplete at many points, from a single-celled green alga to a grass plant, its head bent by the weight of its seeds.

Back now to the quiet hillside in Turkey. The sun has just set, so the leaves are quiet; there's no light to activate their chlorophyll. The flowers are closed for the night; their pollen will be carried by the next day's wind where it may, and the water moves around through the plants' xylem, all in all anticipating another good day tomorrow. We don't have the ears to communicate with these grasses, and it's too bad, because if they could talk and we could understand, they would have quite the story to tell.

CHAPTER TWO

The First Encounter

All the while that the plant kingdom was morphing, growing, and evolving, the animal kingdom was on its own circuitous trajectory. When green algae first washed up on land, the oceans had been teeming with life. But plants weren't the first higher organism on terra firma—a centipede, maybe 18 inches long with a couple of dozen legs, had left its imprint as it scuttled across the sand dunes tens of millions of years before the first liverwort, up on land to lay eggs or escape a predator. But land was no more hospitable for animals than it was for plants, and without the plants' ingenious cell walls, animals needed a way to hold themselves up outside the buoyant oceans and to keep from drying out. Amphibians were the first group after insects and scorpions to try life outside the pool, and they split their time between aquatic and terrestrial environments. Early amphibians were covered with scales, like the later evolving reptiles, to maintain their moisture on land. Mammals didn't appear and flourish until long after plants were flowering, but our tale focuses on just one section of their part of the evolutionary tree: the branches that ended up in *Homo sapiens sapiens*.

The details of how our species evolved are pieced together from scattered skeletons and bones, and each new discovery not only adds another piece to the puzzle but sometimes has us revising much of the story.[1] That said, the rough outline is that 15 or 20 hominin species proliferated over the last 6 million years. Our genus, *Homo*, split off from the others about 2 million years ago, encompassing six or eight other species, depending on how you interpret the fossilized bones. Those 2 million years were a time of major climate instability, with glacial periods lasting about 100,000 years, interspersed with warm interglacial periods lasting about 10,000 years. With major climate swings, plants die back in one area as the weather patterns become inhospitable, and colonize a new area. Animals follow the vegetation. If you could have watched all this with time-lapse photography taken high above the earth's surface, it would look as though the planet were breathing, or maybe breathing while doing yoga on a geologic time frame. The long exhalation is the expansion of white sheets of ice, extending from the poles and the mountaintops, while green forests contract toward the equator. At the same time, the blue bodies of water that make up the oceans and lakes recede, as that water gets locked up in glaciers. The inhalation happens when the climate warms, ice sheets shrink, sea levels rise, and plant and animal species migrate back toward the poles and higher elevations. It's a very fluid view of life, with populations expanding or contracting with shifting environmental conditions. The pace and intensity of these changes weren't evenly distributed across the globe. Temperature changes were more extreme toward the poles, and rainfall patterns shifted with changes in the circulation of air masses.

With all the migrations in response to climate change, plant and animal populations could become isolated in small pockets for long periods, especially when the climate was more restrictive. Those two conditions, small populations and isolation from the rest of the species, are perfect conditions for rapid evolutionary change. Small populations can change faster than larger ones, just because it takes fewer generations for an advantageous variant to spread through a small population. Or by chance, the isolated population may include just a subset of the variation in the whole species, making that group unique without any specific selective force. It would be like asking for a few volunteers from a crowded auditorium, and by chance all those volunteers were red-green color blind. Small, isolated populations can rapidly differentiate from one another, hastening the process of speciation.

For our lineage, one of the greatest changes that occurred during those 2 million years was the development of our brains. If you track the size of hominin skulls over time, brain size stayed roughly the same between 7 million and 2 million years ago as that of our great-ape ancestors. The increase happens in three steps: the first about 2 million years ago, another jump in size about a million years after that, and the last one about 200,000 years ago. The result is that now our brains are about 3 times the size of that of the great apes. If there was a specific trigger for the increase in brain size each of those times, we've not yet figured out what it was, but these seeming leaps aren't unusual for the evolutionary record, which is characteristically sparse on details.

Adaptations related to a body part (like the brain) or a chemical compound are often retrofitted to something quite useful long after their origin. Lignin in the plant cell wall is like that. It started out as a molecule that helped deflect UV radiation in algae, but later got modified and put to use for structural support in plant stems. Feathers are another example. They are beautifully engineered for flight, but had first appeared on dinosaurs long before the evolution of birds and any possibility of their taking to the skies. Similarly, the human brain grew larger over time, and we have put that to use in helping us manage social situations and adapt to changing climates, but the specific triggers that resulted in our ancestors' increased brain size remain a mystery. One thing that's not a mystery is that big brains are an energy drain. Accounting for just 2% of our body weight, our brains use 20% of our metabolic energy. It takes a lot of glucose and oxygen to feed the neurons that are firing as you read these words and to maintain all your established neural pathways.

That's why one of the main problems occupying our brains is where to get food.

What early humans ate is a key part of the story of wheat, because it's not intuitively obvious why they would choose to eat grass seeds in the same way they might choose to sample wild grapes. Wild fruits may be small and seed filled, bitter or sweet, and clearly often unappealing. But grass seeds are small and hard and impenetrable. The food choices of our hunter-gatherer ancestors are a topic of great interest, not only to anthropologists but also to people trying to come up with a more thoughtful approach to eating. Some of the most popular modern-day diets start with the underlying assumption that if our food choices follow those of our ancestors, we can avoid obesity and food-induced health problems. Whether it's the paleo diet, the raw food diet, or a vegan approach to food choice, the idea is that hunter-gatherer food choices were more natural, and in that way better for us. That's debatable given the differences in lifestyle and caloric output, but the quest to return to a more natural life is one way to address our dissatisfaction with today's stresses.

To reconstruct the diets of early humans, anthropologists extrapolate from scattered pieces of evidence, including bone chemistry of fossilized remains, the anatomy of our ancestors' jaws and the size of their teeth, and any food remains near their campsites, along with the likely plants and animals in the area that were potential food sources. The plant component of our diet is firmly ensconced in our history, beginning with our early hominin ancestors that ate leaves, roots, and nuts. Plants are a problematic food choice; they aren't easy to eat. Remember those cell walls that provide physical support for plants, once they moved out of the buoyant comfort of their ancestral aquatic home? The cellulose microfibrils that make up the walls' structure are indigestible for animals. This is the roughage that's good for us, because it moves through our digestive system without breaking down. That's great if the rest of your diet is nutritious and you're worried about bowel function, but not so great if you're nutrient starved. Cows and other grazers have a specialized stomach that houses a bacterial community which can break down the cellulose, but in our guts, not a single one of the hundreds of species or trillions of individual bacteria that live there can do the same.

Neither entirely vegetarian nor vegan, the diet of our early hominid ancestors likely resembled that of the great apes in that it included insects and small amounts of meat. Our culinary practices started to deviate as our brains grew larger. But while the brain case was increasing, the thoracic cavity of the hominin skeleton was decreasing, suggesting a smaller gut size. The combination of the metabolic demands of a larger brain—we need to increase our energy intake—and the decrease in digestive tract size points to a transition to a higher-quality diet. And this is where meat comes in. Anthropologists have suggested that our shift to eating meat, first as scavengers, later as hunters, was the prerequisite for large brain development. The oldest spears come from an archaeological site 500,000 years old, so hunting is at least that old. But raw meat is far less digestible than cooked meat, meaning less energy from the same amount of food. The same goes for the digestibility of starchy roots—things like yams and cattail roots. These plants and others in the lily family, the squash family, and the potato family have thick roots that function as underground storage organs. In the fall, plants move carbohydrates from the leaves to the roots, saving them for the following spring, when the same resources can be moved aboveground to give the plants a head start as soon as the weather warms. With their big leaves, these plants are easy to find, so they were a reliable source of food for our ancestors. But the nutrients are much more difficult to access if eaten raw.

Evidence of our ancestors using fire comes from 1.7 million years ago. But using fire occasionally, for example after a lightning-strike wildfire, is different from either starting or maintaining a flame for regular use. Campsites with hearths that were used continuously appear about 400,000 years ago. The changes in food choices that followed our ability to manage fire, along with physiological and morphological changes in our digestive system, combined to bring about shifts in our diet that further separated us from our great ape ancestors. With all aspects of the human evolution story, however, each archaeological study adds to the entire account, and our understanding in 10 years could be different from what we know now.[2]

Early humans were hunters and gatherers, moving seasonally with the availability of food. Their shift from a nomadic lifestyle, in which they relied on hunting and foraging wild species, to a sedentary lifestyle was not always precipitated by the advent of agriculture. But in those cases where hunters and gatherers formed settlements, they often eventually made the transition to growing crops and raising animals for food. That transition was followed by a rapid increase in population, due to an increased efficiency in providing enough food and a higher reproductive rate. When we started to farm and what we grew varied around the globe. Depending on how you interpret archaeological evidence and what assumptions you make about people migrating with crops, there are somewhere between 10 and 20 regions where people independently initiated agriculture and domesticated crop plants. The earliest farming likely occurred in the Near East, with cereals and lentils the first crops grown. In South Asia and East Asia, two other grasses were the first crops—rice and millet. Root crops—taro and yam—along with bananas were the first agricultural species in New Guinea; squash, corn, and avocados were domesticated in Mesoamerica; peanuts, squash, beans, and potatoes were grown in South America; and in North

America agriculture started with squash and sunflowers, 8,000 years after wheat was first cultivated in the Near East.

Domestication happens when humans knowingly or inadvertently select traits in a plant or animal that would make it more useful. Generally, domesticated plants are bigger and hardier, the result of our having chosen the plant or seed that looks the best for preparing food and for planting. Other characteristics of domesticated plants are selected by chance as the result of our planting and tending. Consider, for example, that wild grasses have a brittle attachment point between the stem and the seed; when the seed is ripe, it falls to the ground, optimally on a site favorable for its germination. Ecologically, it's tough to argue with that strategy. Humans, however, have a hard time gathering grass seeds by picking up individual grains from the ground, so the seeds most likely to be collected are the ones that are still on the plant. A single gene causes the attachment point to stay strong, so if both brittle and strong attachment variants are present in the wild population, those seeds that are easiest to harvest are collected and planted again, eventually giving rise to plants with traits that differ more and more from their wild ancestors.

The beginning of agriculture is not easily detected in the fossil records. Wooden tools and plowed fields rarely leave any trace. But the morphological changes that come with plant domestication is actually visible in the organic fragments found in archaeological sites: larger seeds and a breakpoint at the base of the seed that is either smooth (as when a wild plant drops its seeds) or jagged (as when seeds are threshed off the stem). We also know that agriculture was likely happening even before we find a change in plant traits, by the increase in weedy species common in cultivated fields but not in undisturbed areas. The earliest agricultural sites are in the Levant, an area east of the Mediterranean Sea, bordered by the mountains of Turkey and Syria to the north and deserts to the east. To visit one of the sites containing some of the earliest evidence of domesticated grains, we must travel across time—back about 13,000 years—and cross the Atlantic Ocean, pass through the Strait of Gibraltar, and journey over the Mediterranean to a formerly quiet land now called Syria. We look for the Euphrates, the longest river in the region, stretching almost 2,000 miles from the high mountains of eastern Turkey to the Persian Gulf. Our stopping point is at the river's elbow turn a long three days' walk east of Aleppo. Here, on a rise above the river basin, lies the site that will become the settlement of Abu Hureyra.

Location, Location, Location is the mantra of real estate sales, and the same holds true for finding a good site for settlements 13,000 years ago. At the site that will become Abu Hureyra, your eyes can follow the Euphrates upriver to the northwest for more than 30 miles, and downriver to the east for another 10. The river at this time is wide and fast running, especially when the spring snowmelt, carrying rich red-brown soil from the Zagros Mountains, spills over its banks. Blanketing the shoulders of the river, bright-green reeds and cattails stand in water, with bitterns nesting and herons hunting amid the tall green leaves. Thickets of willows and tamarisks break up the low-lying marshes, and slow the force of the rushing river water. Moving away from the river, a band of tall forest, composed of poplar and ash trees, creates a respite each summer from the persistent heat. Plantane trees tower above everything, with trunks so wide it takes two people to reach around one. The smooth trunks are crisscrossed with vines, lavender clematis flowers, and clusters of small purple grapes, like an Impressionist painting of a verdant oasis.

Above the river plain, on the shelf where Abu Hureyra lies and beyond, trees and shrubs dot the grasslands, stretching as far as you can see. Red flowering terebinth, with a turpentine-like sap, and oaks are the main plants here, but you can also find wild pear, almonds, cherries, wolfberries, capers, service berries, buckthorns, honeysuckles, sumac, and hackberry. And where there aren't shrubs, grasses grow in thick stands, including the wild cereals that will be the first plants that villagers cultivate. Plants from the pea family—lentils and wild vetch and clover—grow mixed in with the grasses. These plants are the ancestors of our pulse crops, the peas and lentils and beans.

Behind you is the open grassland steppe, gently undulating open fields with scattered trees and bushes. It's green every spring and dotted with flowers, but much of the year it's a dry, golden grassland. Oak trees, some tall and upright, others more like spreading shrubs, punctuate the view, providing patches of contrast to the open plains. And cutting through the fields are the wadis, fresh water flowing in ravines carved by seasonal runoff, lined with almond shrubs, wild bitter watermelon, and thick stands of grasses.

This idyllic spot had been used by humans even before its colonization 13,000 years ago, and was occupied for a long time afterward. It was located high enough above the fast-flowing Euphrates to be outside the flood zone, yet adjacent to a fertile plain, where the river deposited rich sediments every spring. But if you traveled to it now, which for political reasons is unlikely, the site is at the bottom of Lake Assad, Syria's largest lake, created in 1973 by the construction of Tabqa Dam. And the areas adjacent to Lake Assad don't exactly match the description of Abu Hureyra from 13,000 years ago. The climate in the Middle East was wetter then, with enough spring and summer rainfall that trees grew in the grassland steppe, creating an open forested parkland in an area that nowadays is parched plains.

What we know about Abu Hureyra comes from an extensive archaeological excavation begun in 1971 and ending abruptly in 1973, when the dam was completed and the site flooded.[3] While excavating, archaeologists collected fragments of organic materials by floating all the soil obtained from the hearths in large tubs of water. The soil settled to the bottom of the tubs, while the light pieces of seeds and plants floated to the top. Those organic samples were collected and analyzed by the archaeologists over the next several decades, with new technologies applied as they developed to more accurately date and identify fragmented remains. From the excavation, we know that the site was colonized about 13,000 years ago for about 1,500 years. Then there's not much record of human activity until about 10,500 years ago, at which point the site was colonized again for another 2,500 years.

The people who lived at this site were smaller and more muscular than most of us. The women were just a little over 5 feet tall, and men were about 5 feet 4 inches tall. Abu Hureyra was settled tens of thousands of years after the animals were painted on the walls of caves in southern France, and after women were carved in ivory in southwestern Germany. The people who lived in the Levant during this time had brains that worked like ours; they fell in love; they had kids; they grew old and died. But all that we have left of their lives are traces of their material culture. We know that they lived in circular huts, with floors about 2 feet below the ground's surface and 3 or 4 yards across. Posts around each hut's outer edge supported a roof of grass thatch or perhaps hides. People added rooms to their dwelling by digging out another pit; smaller rooms were for storage and larger rooms for living. The hearth was placed outside the hut, and the abundance of tools along with the plant and animal remains left near the fire provides clues as to how the Abu Hureyrans lived and ate.

The early settlers of Abu Hureyra weren't farmers. Indeed, the ecological richness of the site meant that people could hunt and gather from the local surroundings year round. Archaeologists estimate that over 120 species of plants at the site could have been a source of food for Abu Hureyrans. In the spring and early summer, wild wheats, ryes, and other grasses could be harvested for their seeds, which were then ground to make a paste or porridge or flatbread. Wild lentils and other seeds from the pea family would be ready to harvest throughout the spring and later in the summer. And seeds from both the grass family and the pea family could be stored for later use. Wild fruits and nuts would ripen in the late summer, and buds and leafy greens were available throughout the spring and summer. The list of fruit and berries that likely grew at this site sounds like a scan of the inventory at a boutique grocery store: wild grapes, figs, pears, hackberries, mahaleb cherries, sour wild plums, yellow hawthorn, wild capers, juniper berries, and more, with the cautionary note that wild fruits aren't the same size or sweetness as our domestic varieties. In the fall and winter months, the starchy roots from wetland plants, including bulrushes and club rushes, could be dug up from the marshy areas next to the Euphrates; these could be pounded and soaked and eaten.

The evidence for what people of Abu Hureyra actually ate is partly based on bits of seeds found in the soil near the settlement's hearths. Charring grass seeds made the husks easier to remove, and charred remains are more likely to be preserved. And botanists can reconstruct plant communities around the site for that period, based on their understanding of climatic conditions at the time and their knowledge of the distribution of plants in undisturbed parts of the Levant. Using that work and borrowing from anthropological studies of the diet of modern hunter-gatherer cultures enable us to piece together the likely diet.

For meat, villagers hunted gazelles that migrated toward the moist grasslands of the Syrian valley, probably in large herds, just after their young were born. The timing of the hunt is indicated by the age of animals' teeth and bones found near the settlement fireplaces. After a large hunt, it's likely that the whole village was involved in drying and preserving the meat in the ensuing weeks. Tools for scraping and needles and awls for perforating hides suggest that villagers used the animal skins, perhaps hanging them on the walls of their dwelling for insulation or using them as a roof. Onagers, a wild ass, were also hunted around the same time of year, and boars, hares, foxes, and a rich assortment of birds lived in the forests near the village and were hunted as well.

Several hundred years into the first period of settlement at Abu Hureyra, the climate changed rapidly, becoming both cooler and drier. The Younger Dryas, as this period was named, happened across the Northern Hemisphere, and was marked first by a rapid cooling and then a rapid warming 1,200 years later. The Levant received less rainfall in the spring and summer during this period, with more moisture bound in snowfields in the mountains. For a people who depended on local plants as their main food source, the disruption caused by changes in climate was huge. It's like the grocery store moved thirty miles away, and, oh yeah, no car. Their change in diet during the Younger Dryas is reflected in the plant fragments found in the hearths of Abu Hureyra. Trees and plants from the forested habitats disappeared from those remains, replaced by roots and seeds from more drought-tolerant plants. Because wheat and rye probably grew among the patches of trees in the oak woodland, as trees declined with the decrease in moisture, the grasses would have declined too. And that's reflected in the decrease in wheat and rye seeds found near the hearths within just a few centuries after the climate started cooling.

What appears around the same time is perhaps the first evidence of agriculture—domesticated rye grains and weeds that grow in cultivated fields. I say perhaps, because not everyone agrees on how to interpret this fossil record. One interpretation is that while wild stands of rye and wheat declined, the people of Abu Hureyra started cultivating their own cereals, or at least rye. This would be the earliest documented evidence of agriculture, about 12,000 years ago. Besides a steady increase in weedy species at the site, there is an increase in the seeds of two plants from the pea family—lentils and wild vetch—which likely would not have grown near the site with the change in climate, raising the possibility that those species, too, were being farmed.

While the existence of farming during the first settlement period of Abu Hureyra can be disputed by scholars, the presence of this activity is unequivocal during the second settlement period, beginning after the Younger Dryas when the climate warmed again. It's not clear whether people were living at the site in the almost 1,000 years between the two settlement periods. Domesticated wheat and rye seeds from that period have been found, but not a lot of other things. The similarity in tools between the earlier and later periods suggests a cultural stability over time, what you would expect if the same families stayed there, maybe living nearby in a site that was not part of the archaeological excavation. Although the village during the first settlement period was home to several hundred people, the shift from hunting and gathering to agriculture supported a larger population. As many as 2,500 people lived in Abu Hureyra during the early part of the second period of settlement, and the population eventually reached a peak of 6,000. Instead of circular thatch-style huts, dwellings were now made of mud bricks, clay mixed with straw and pebbles, then dried in the sun and bound together with mud mortar. Houses were rectangular, with painted plaster walls and floors, and hearths were shared between several houses. The village was large, which was not so unusual for the Levant at that time: at least a half dozen other villages in the region spread over 20 to 40 acres. People moved from smaller villages, likely because of environmental degradation, and pooled up in larger ones. The first written records date to 5,000 years later, so we can only speculate on how the people of the Levant lived, based on what they had left behind that survived over the intervening thousands of years.

Initially, people in the second settlement of Abu Hureyra farmed and gathered wild plants and hunted gazelles. In addition to rye, they grew three kinds of wheat, along with barley, lentils, peas, vetch, field beans, and chickpeas. Archaeologists found no agricultural tools at the site, so the villagers probably broke the soil with wooden tools. The fields were probably cleared in the late fall, after the first rains provided enough moisture for the soil to be easily worked. Seeding of the cereals also likely occurred in the fall, followed by weeding between the young plants. Work in the fields decreased for the winter, with the cool temperatures slowing the growth of the crops and the weeds. As with modern farm communities, this was a time for rest and leisure. As the weather warmed in the spring, villagers would plant peas and beans. The gazelles returned in April during their annual migration, and hunting and processing the meat presumably involved most of the settlement. After that, cereal harvest would start toward the end of May and into June, along with foraging for wild plants. The lentils and peas and beans would be harvested starting in July, and those crops then would be processed and stored.

Any community populated by thousands of people needs some kind of political structure to establish rules for living together and to help resolve conflicts. But at Abu Hureyra, there is no evidence for a hierarchically structured community, like you would expect if there was a single ruler. The houses were all about the same size, and burial practices were consistent, with no signs of more elaborate rites for some people. There's also no evidence of fighting, either weapons or, in human skeletons, crushed skulls or broken bones indicative of death from a physical attack. The people of Abu Hureyra were perhaps a peaceful and hardworking people.

The village was abandoned a little less than 8,000 years ago, as the climate in the Euphrates River valley got drier. We know that the impacts of the growing population on the surrounding area had reduced the potential for hunting and gathering, as all evidence of reliance on wild plant species disappeared for the last years of settlement. For meat, the villagers shifted to domestic animals—sheep and goats. But we don't know where the people of Abu Hureyra went or what they did. There's no sign of a new village arising nearby; and with the changes in the climate, the village was surrounded by increasingly arid lands.

The beginning of agriculture in the Levant was made possible by its fertile soil—at Abu Hureyra, the soil deposited by the Euphrates during the spring flooding—and its warm and moist climate at the time. People in the Levant had settled in villages before they ever farmed, and cultivation was perhaps what allowed them to remain in their villages after wild plant and animal populations declined, caused by either overuse or climate change. The most satisfying way to conceptualize the origins of agriculture is that of a birthplace in the Fertile Crescent, with crops and cultivation methods radiating through time across the continents. Correspondingly, the best ending to the story of Abu Hureyra is that even though life stopped in this little village by the Euphrates, its inhabitants' way of life moved on to a new site, somewhere to the north or perhaps to the east. But in fact, we don't know whether people moved between sites, carrying the knowledge of agricultural practices with them, or whether different villages began farming independently. We know that on a global scale, agriculture arose independently in a number of areas, and methods of farming and animal domestication varied from one spot to another. Although scholars argue extensively about what might have been the tipping point causing our ancestors to start cultivating plants, it's not surprising that our ancestors worked out the details for planting, tending, and harvesting their crops. For people who forage for plants, young plants, such as sprouts, are a sure source of food. Sometimes a seed is still attached to the young seedling. Understanding the relationship between a seed, a young plant, and a mature plant seems pretty straightforward. And it requires just a little bit of experimenting to figure out how to plant seeds. You find young plants only in moist soils, so adding water and getting rid of nearby plants that are crowding out the newly planted ones also seem straightforward. It's a logical step, then, for people who are already settled in one place to make the transition from foraging to farming. Thus, the most unlikely aspect of this story is perhaps how, many generations before we ever planted seeds, our ancestors started using grass seeds as a food source in the first place.

Seeds are the plant's future. If a plant can make a good seed, it has a higher likelihood of lasting another generation. And a good seed is one that can wait in the soil for favorable growing conditions. When there's enough moisture, warmth, and sunlight, the embryo inside the seed stretches out, sending the first roots down to the soil and unfurling the leaves, getting everything working so that the seedling can photosynthesize on its own. Besides a protective outer coat, the seed is composed of the embryo, packed with all the essential nutrients, and the storage tissue that fills most of the seed with the energy reserves for early growth. All this is exactly what makes the seed a good source of food for animals and us. Grasses are packed with both carbohydrates and proteins, the quick-burning and slow-burning fuel we need and crave.

The other aspect of being a good seed is to protect all your goods from creatures craving food, and plants do that with chemical and physical means. Cyanide protects apple seeds, and capsaicin protects pepper seeds. Other plants create a shell hard enough to discourage most predators. And then there are plants like peaches and almonds that use both means. The wild almonds around Abu Hureyra, for example, had to be detoxified to remove cyanogenic glucosides. If they weren't fermented or roasted before eating, a snack of a couple dozen almonds would have been a fatal dose.[4]

Grass seeds have the advantage, from the perspective of a

hungry forager, of not needing detoxification. Their protection for the embryo is physical, and begins with a pair of bracts, modified leaves that botanists call glumes, and we more commonly refer to as hulls. Inside the hull are another couple of layers, the fruit coat and the seed coat, which have fused to make a hard layer, which we know of as bran after it's ground. Bran is another herbivore deterrent, being primarily indigestible (and thus a great source of fiber for us). Inside the bran are the embryo and the storage tissue. The embryo is the baby plant from wheat's perspective, what becomes the wheat germ for us. It's got the highest nutrient content. The storage tissue surrounds the embryo and makes up the majority of the seed. It's the storage tissue that supplies the developing embryo with carbohydrates until it can photosynthesize on its own, and it's what we use for producing white flour.

So pause for a moment, and imagine being a hungry forager 13,000 years ago. You find a stand of grasses, their heads bent with the weight of their seeds. The ripe seeds fall off when you shake the grasses, so it's easy to gather a handful. But if you try biting into one, you'll break a tooth. With a big rock you can break the hull, but the hull is thick, definitely thicker in wheat than other grasses, and the whole seed sometimes breaks in the process. It takes a few blows with a heavy rock to remove the hull. What you'll discover later is that if you soak the seeds first, or if you char them over a fire, the hull is easier to break. But for now, you have pounded the wheat seeds with a big rock, resulting in a mix of tough fragments and seeds. With a shallow basket, you toss the whole lot into the air so a light breeze can carry the husks away, leaving you just the heavy seeds. But no, this is wheat—the hulls are thick, and the whole mess lands back in your basket. After losing patience with picking out the seeds by hand, you decide to try separating them out by holding the basket at an angle, shaking lightly so that the heavier seeds slide a little further forward than the broken shaft. Finally, you've got a small pile of seeds. Try biting one now, though, and you'll break another tooth.

With the big rock you used to break off the husk, pound the seeds until they break into little pieces. That leaves you with pieces nonetheless just as hard as the whole seeds. Next, try soaking the seeds in water. Now you have a mush that satisfies your complaining stomach, but still has some really hard pieces in it. In fact, one of the indications that our Neolithic ancestors were eating wheat is the cracks and pitted areas in the teeth of their skeletal remains. To make the wheat seeds more edible, then, grinding is better. The oldest tools for that purpose date to about 30,000 years ago, at sites across Italy and eastern Europe. At Abu Hureyra, archaeologists found querns, two-part grinding tools. The lower piece of rock, usually made of basalt, was smoothed into a flat or saddle-shaped surface, and then a rubbing stone, a rounded rock about the size of a jam jar, was held for crushing. From the time of Abu Hureyra up until the fifth century BC, saddle querns were the state of the art for grinding seeds.

Saddle querns weren't especially large, maybe a foot and a half long and a foot wide. To efficiently grind the seeds, you kneel behind the quern, and spread a thin layer of seeds across the base rock. If you have too many seeds, unground pieces can get lost within the thick layer of ground ones. By leveraging the weight of your body, you can crush the seeds into flour in just ten passes if you're lucky. Then you wipe the flour off the surface stone, and add the next layer of seeds.[5] From the skeletal remains found at Abu Hureyra, it's clear that most of the grinding had been done by women. The effects of grinding grass seeds are indicated in their big toes, hip and knee joints, and shoulders, where the repeated action of pivoting their weight resulted in curved bones or bones that were buttressed at the joints. Wheat seeds aren't the easiest seeds to prepare for eating, but they store well. Ground flour will go rancid, but if only as much is ground each day as is needed, the stored seeds are a dependable source of food for months.

The breakthrough for cereal eating was the advent of pottery, in which wheat porridge could be cooked. Before that, people had eaten wheat either as roasted whole grains or in a ground form mixed with water to form a paste, and then eaten raw or cooked as a pancake or in some form of bread. With pottery, whole or crushed grains could be boiled until they were soft. Earthenware was introduced at Abu Hureyra late in the second period of settlement, a little more than 8,000 years ago. That coincides with the time that the population grew to its maximum, and the new cooking method was easier on people's teeth. After the introduction of pottery, the teeth found at the site were in much better shape.

The wild wheat that grew near Abu Hureyra and that our ancestors first cultivated were two species—einkorn and emmer. After a few thousand years of being planted and harvested and ground, einkorn was mostly discounted, relegated to the less desirable foods. It grows better than the other wheats on less fertile soils, but the seeds grind a little coarse, so it's mostly deemed suitable feed for animals. But with our recent interest in eating how our ancestors' ancestors ate, einkorn is grown now as a specialty ancient grain that's made into cookies and crackers for foodies.

Emmer fared better than einkorn, maintaining a respectable level of prominence for thousands of years longer primarily because it could grow well in hotter climates, where einkorn withered. Emmer wheat was particularly honored by the Egyptians, finding its way into the tombs of the pharaohs as an accompaniment into their afterlife. Nowadays, emmer has a similar status to einkorn, as an ancient grain for specialty markets. Emmer also left its mark in another way, having spun off another subspecies: durum wheat. Durum wheat has hard seeds characteristic of both emmer and einkorn. Those seeds grind up into a flour that makes a dough that can be rolled out into thin layers, cut into shapes and dried, and then boiled to make a perfect dish of pasta.

But overshadowing both emmer and einkorn is bread wheat. The first farmers didn't grow bread wheat, because the species had yet to exist. The origin story for the different wheats can be deciphered in part by looking at their chromosomes. Einkorn wheat is the oldest of these species and has 14 chromosomes, two copies of each of seven chromosomes. A long time ago, maybe 500,000 years, a close relative of einkorn was fertilized by a pollen grain from a weedy goatgrass, a species that doesn't have particularly interesting characteristics as far as food goes. Hybridizations between two different species aren't supposed to happen; we define species as a group that's reproductively isolated—in other words, they don't hybridize with other species. But plants in general and grasses in particular are rule benders.

Emmer is the result of the hybridization between einkorn's relative and a species of goatgrass. We know that because emmer wheat has 28 chromosomes: 14 from the einkorn parent, and 14 from the goatgrass parent. Despite the differences in chromosome number, both einkorn and emmer species have seeds that are larger than most other grasses, with tough hulls that protect the seeds. They are a good protein source, but one that required a fair amount of effort to clean and prepare them as food.[6]

Bread wheat features a trait that made it a more appealing food source than the other wheats. It likely originated in a farmer's field of emmer about 8,000 years ago. Apparently, one of the emmer plants was fertilized by a pollen grain from yet another species of goatgrass. That cross happened without any human help, and that hybrid—our bread wheat—has the 28 chromosomes of emmer (14 from einkorn's relative and 14 from a weedy goatgrass) and now an additional 14 chromosomes from a different species of goatgrass. That's a total of 42 chromosomes from three different species. The last set of 14 chromosomes conferred on the new species both a softer seed and a softer hull that easily breaks off the grain. The softer hull eliminated the need for charring or soaking and pounding, and the step of separating the hull from the grain; the softer seed meant that grinding took a fraction of the time. This awkward hybrid, bearing a bulky set of 42 chromosomes, became the most appealing of the wheat family. Besides the ease of preparing the seeds, the proteins in the bread wheat grains form a network of strands that give a light consistency to the dough. For people developing cuisines based on breads, the lightness of the bread made with this wheat above all others was key for the rising prominence of bread wheat. Today, bread wheat accounts for about 95% of the 700 million tons of wheat we grow.

Although we will never know what became of the people who abandoned the village of Abu Hureyra, the seeds of einkorn, emmer, and bread wheat weren't a casualty of their departure. Their spread and our spread throughout the Middle East, through the Balkans, and up the Danube River take us to where our story continues.

CHAPTER THREE

Intertwined Lives

For a plant, life is pretty straightforward. Lacking the capacity to anticipate and worry, all that plants respond to is their external environment. Unlike animals, plants can't run away, they can't move to a more comfortable spot, they can't avoid a competitor by finding new hunting grounds. Instead, they respond to whatever the world throws at them, or they die. The great evolutionary sieve lets pass only those that germinate, grow, mature, and then produce the seeds for the next generation.

Unfortunately, success doesn't necessarily beget success. Species, even the seemingly most successful ones, still must manage the inevitable changes—wet years or dry years, warming or cooling trends, new neighbors adept at taking up nutrients, smarter predators, wildfires, and on and on. A necessary part of the evolutionary gambit is the option to change, to be flexible in the face of an unpredictable environment.

There are a handful of examples where a plant's or an animal's response to environmental variation is rapid enough to be observed either within our life span or with the help of successive naturalists taking notes bridging several generations. The peppered moth is one of those examples, a little black-and-white moth that, like most other moths, flies at night and rests during the day. At night, peppered moths must avoid other nocturnal animals like bats, and during the day they rest hidden from birds and other predators who hunt with their eyes. In forests with white-barked birch trees, the lighter-colored moths are camouflaged from day-feeding birds, and so have a higher survival rate than their dark-colored counterparts. Now, bring on the Industrial Revolution, with factories that spewed forth clouds of coal-generated soot. The trees' white bark was covered with a layer of dark soot, leaving the white moths vulnerable to hungry daytime predators. With that, dark moths survived at a higher rate, and the population shifted to better-camouflaged dark moths. Decades later, with greater awareness of the impacts of pollution, England established restrictions on industrial emissions. The soot level declined, and the birch trees stayed white. With those changes, the moth populations shifted back to a preponderance of white moths, because black moths were now more visible to predators.

The evolutionary sieve in this case is the predation on the moths; the superiority of the dark versus light forms of the moth is context dependent. The capacity to change in response to environmental fluctuations is key for a species' evolutionary longevity. Indeed, flexible responses are the underlying secret to most of life's successes—family relationships, long-term business profits, and personal happiness. When we think about growing food for this year and into the future, we would like our crop plants to respond with alacrity to whatever environmental variability occurs, whether that is climate fluctuations, fungal pathogens, insect outbreaks, or less than optimal soil conditions. At the same time, we would like our crops to grow a lot of food. We're asking for a lot: flexibility to change and optimal performance. We achieve optimal crop productivity by eliminating all but the most vigorous plant varieties, so that every plant in the field produces as much food as possible. That practice is in direct conflict with promoting variation within the crop species, the variation necessary to respond to environmental changes. This paradox—breeding for the best variety while conserving variation—is the major challenge for managing our crops for our children and our children's children. But let's take this story back to where it left off, in the early days of agriculture when considering the next meal took precedence over the meals of future generations.

For wheat, the onset of agriculture changed the rules of the game. Wheat plants were now being tended, their seeds carefully placed in the soil and in some locations provided with water when needed, with competitors removed and seeds saved for planting the following year.[1] Humans had become a dominant selective force for the wheat species. With the peppered moths, increased predation by birds on the moths that stood out against the forest background led to changed colors in the surviving population; in the case of wheat, when our ancestors planted the seeds from parent plants that germinated and grew the best in the local climate, yielding plants that were easiest to harvest, they changed the traits of domesticated wheat. Once humans became part of the evolutionary sieve, the traits they selected could in fact conflict with those beneficial for wild wheat. The dense hull that surrounds wild wheat seeds protects the embryo and its nutrient source from hungry predators. But after a variant showed up in the population, caused by a single genetic mutation resulting in thin hulls that easily broke away from the rest of the grain, our ancestors started planting seeds from plants with that trait. The thin hulls that left seeds vulnerable to attack were much easier for our ancestors to use, eliminating the step of pounding the seeds prior to grinding. And wild emmer wheat's insurance policy of producing two seeds per spikelet—one that germinates the first year and another that germinates the next—wasn't very useful for our species, who wanted all the seeds to germinate soon after planting.

How the first farmers turned over the soils and planted and harvested wheat is obscured by the layers of life that accumulated on top of any record they might have left behind. But we know that the people of Abu Hureyra were part of a larger pattern of settlement in Southwest Asia. Across the region were a number of villages the size of 15 or 20 football fields, their rectangular, mud-brick dwellings packed close together, home to thousands of people. Archaeologists can sift through the sediments for imprints of buildings and abandoned tools, along with charred sediments revealing which plants had been used as food. Without written documentation of how the people lived, however, our understanding of how societies were organized throughout the first millennia of agriculture is largely speculative. But clearly, concentrating settlements into larger areas necessarily changed the economy of scale. In a small village, families could have their own farm plot near where they lived. As villages grew, limited access to good farmland likely meant that farming efforts were collective. The focus for food production shifted from a concern to gather and process enough food for one's family to producing food for a much larger group, including people who may not be related but reside in the same village. Instead of the grain-processing arrangement used at Abu Hureyra, where each house had a grinding area, some villages built centralized grinding rooms housing multiple querns. A village's collective production of food changes its social structure, and the details of those changes through time will be the focus of scholarly work for years to come.

Not everyone alive at the time of Abu Hureyra lived in villages, however. The geography of the Levant is a patchwork of Mediterranean coastline, coastal plains, steep mountains, and rift valleys. Add the floodplains and riverine forests of the Tigris and Euphrates Rivers along with the dry deserts, and you have an area uniquely situated for a mixture of lifestyles. In areas that were too dry for farming because of sporadic rainfall, or in regions without good soil, nomadic herders moved with the seasonal availability of water and grazing areas. As the climate fluctuated between years or decades, gathering areas for certain foods shifted as plants died back in places that now lacked the right conditions for growth, and sometimes expanded into new areas.

The mix of lifestyles supported trade activity between communities. Herders traded with villagers—grain, animals and animal skins, cloth, and tools—to supplement villagers' dietary and other needs in exchange for grains from the settlement that could be carried easily. Yet trading wasn't a new development. Obsidian, a glassy black rock, had been traded by hunters and gatherers thousands of years earlier; hence trading had been practiced long before the beginning of agriculture. The realization that crops could be a commodity motivated the development of agricultural technologies that would forever change our relationship to food production. Consider, for example, being given the task of finding food in an age without grocery stores or markets or refrigerators. First, imagine foraging or hunting enough food for your own survival. To make it a little easier, you could situate yourself in the rich habitats of the floodplains near Abu Hureyra. Once you're comfortable with that idea, consider finding enough food for you and your extended family, a task that's still fairly doable. Now think a little bigger, about being a supplier for your community. As the number of stomachs for which you're responsible multiplies, so do the potential number of hands that could help, and it's likely that some hierarchies of labor would be established: who oversees what, who gets the tough jobs, and who manages and motivates workers. Then suppose someone were willing to trade for food—something like copper, which you'd never seen before, and which could be used in ways that stone or wood or bone tools could not. The simple act of gathering food has blossomed into commodity production.

Our relationship with food, or anything for that matter, transforms substantially when we think about what we need for survival versus what we could gain by trading it for things that would make our life easier and more comfortable. Our squirrel-like instinct to accumulate and cache our treasures along with a feline-like tendency to relax in a warm spot has been part of the impetus to generate the technologies of agriculture. The combination of a big brain and a social nature has allowed us to generate creative solutions to find and then to produce food, in increasingly efficient ways.

But in a world of finite resources and environmental barriers, there are limits to how much we can produce and accumulate. Restrictions were also present for the villagers of the Levant. The advent of agriculture in the Near East was successful because of the topography, the climate, and the set of species that naturally occur there. But semiarid climates, like those of the Mediterranean region, are fragile environments. The foundation of any land environment is based on plants' unique capacity to extract nutrients from the soil, and convert carbon dioxide (CO_2) from the air into living biomass with a physical structure. The stems and branches and leaves and an equal amount of branching roots belowground hold our ecosystems together physically. When we plow up a field to plant crops, we replace that green protective layer with crop plants that grow for just part of the year, leaving soils exposed to the erosive forces of wind and rain the rest of the time.

The environmental vulnerability arises because soils are living, dynamic systems. In addition to the "dirt" part—broken-up particles of rocks which make up its mineral portion—soil contains an equal part of space, the air- and water-filled pores between mineral pieces, like the spaces between tennis balls in a bucket. Those pores are home to a richly diverse soil community—bacteria, fungi, amorphous amoebas, barely visible flatworms, tiny mites, springtails aptly named for their hop—all smaller than the head of a pin, and all represented by tens of thousands of different species, most of which are not yet named or described. The larger players—worms and insects and beetles and centipedes—break up the remains of dead plants and animals, making food available for the microscopic ones.

The plant roots ultimately fuel the whole belowground community. Roots provide in many ways. First, just as cows eat leaves, roundworms and mites are only a few of the herbivores living in the subterranean snacking jungle, munching on roots. Then there are the sloughed tissues, the fine roots that plants shed after just a few weeks of work; they end up being a food source for the decomposing organisms. And live roots transfer a lot of sugars to the soil, sometimes intentionally and sometimes because of accidental leaks. Plants get rid of compounds they don't want, with their roots serving as a kind of emissions pipe for molecules and compounds that are merely waste material for the plants. Roots also release a banquet of compounds into the soil to signal other roots or other creatures they would like to engage with. And sometimes, roots simply leak. A sharp-toothed herbivore bites into a root, and all those luscious cell contents leak into the soil. The razor-like edges of soil minerals can abrade the roots as they grow through the dense matrix—and more precious cell contents get released into the soil. To reduce that kind of damage, root cells excrete a layer of lubricating gel, a thick carbohydrate layer that is like ripe fruit for soil bacteria. Over time, the layers of soil where plants grow become rich in organic matter, the once living (but now dead) pieces of plant and animal and microbe bodies.

The archaeological record has little information about how the people of the Levant dug up fields to plant seeds, but if they broke up and turned over the soils, the rich, dark organic materials buried in lower layers would have been exposed to surface conditions, speeding up their decomposition. Decomposition of organic matter takes the material down to its elements, freeing up nutrients, which is good for plant growth, and releasing the extra carbon back into the atmosphere in the form of CO_2. In agricultural systems, crops mine the soils of their nutrients, gathering up the phosphorus and nitrogen released from minerals and decomposing organic matter. Then we harvest the most nutrient-rich parts of the plants to use as food. From an ecological perspective, this transfer of nutrients from the soil to the food chain is nothing new: historically, herds of bison or other large grazers converted plant tissue into muscle mass, and plants had to figure out how to manage with that. In many grazed plant communities, plants are long-lived, so even if they are broken off and eaten by grazers, they have an extensive root system that holds the soils intact and stores energy to replace the eaten leaves. And some of the nutrients lost to grazing get returned to the soils through urine and feces left in high concentrations after a herd has grazed on a site for a few days. Another form of returning nutrients back to the soils for reuse occurs when animals die and their bodies decompose on the site. Even if scavengers picked much of the meat off the carcass, the bones are a good source of nutrients. And scavengers die too.

The advent of agriculture introduced two challenges to fragile ecosystems. Clearing land for planting replaced all the plants, including soil-hugging layers of mosses and algae and deep-rooted long-lived plants, with crops that complete their life cycle from seed to harvest over just part of the year. Without long-lived plants and surface-binding crusts, soils blow away and get swept off in water running across the site. On top of that, lacking a return of nutrients via waste products or decomposing bodies, soils lose their fertility and become less supportive of plant growth. Rebuilding soils after farming can happen pretty quickly if not too much of the fertile soils have been lost and if there's enough rain to encourage plants to grow in abandoned fields. But in the Levant, the low rainfall meant that plant growth and soil recovery were slow. When communities in the Levant overused their farm fields, their only option was to abandon those fields and clear a new area.

During the first 2,000 years of farming and herding in Southwest Asia, from about 9500 BC to 7500 BC, the number of people on our planet increased ten-fold. Throughout time and across continents, human populations have grown faster after the transition to agriculture, in part because of easier access to carbohydrates and in part because the energy-intensive process of moving children and households is no longer necessary. Settlements spread throughout much of the Levant, with villages spaced about a day's walk apart. As villages grew and settlement sites filled up, people living along the Tigris and Euphrates Rivers migrated north and west into Anatolia, modern-day Turkey. A perfect site had a reliable source of water and fertile soils, and it looked like where the settlers had come from. The open plains filled up first; it took another 1,500 years before people moved into the forests of western Anatolia. Villagers moved with their einkorn and emmer wheat, barley, peas, broad beans, and lentils. They brought their sheep, goats, pigs, and cattle. In addition to meat, the animals were an important source of milk, which was digestible after boiling or used to make cheese, ghee, or yogurt.

These first forays were the beginning of a migration of people and their food species from the Levant that continued over the next 5,000 years. Around 7500 BC, several thousand years after the advent of cultivation, villages in the Levant were either shrinking or abandoned. Climate change may have been the cause, or maybe overuse of a fragile environment; but likely a combination of both factors was to blame. Farmers moved to Crete, where by 7000 BC people grew bread wheat and raised cattle, pigs, and sheep. The expansion to the north and west over the next millennia followed two different routes, both guided by major bodies of water. Populations expanded along the shores of the Mediterranean Sea. To the north, people migrated along the coast of Italy and up onto the shores of the Adriatic Sea. By 5800 BC their reach extended to the Iberian Peninsula, modern-day Spain and Portugal. Along the southern Mediterranean coast, early farmers settled in North Africa, and moved inland along fertile river valleys. The first farming villages appeared along the Nile River by around 5000 BC; people there grew wheat, barley, and legumes, and raised sheep, goats, and cattle. Further expansion into Africa was slow. Farmers settled in Ethiopia around 500 BC, growing cereals and legumes. The African deserts and the monsoon climates of sub-Saharan Africa were a natural barrier to further expansion on that continent.

A second path of migration followed the Danube River, in two successive waves that moved up through the Balkans to northern Europe. The fertile soils of the Danube and its tributaries were the guide for expansion that started in the Carpathian Basin about 6000 BC. The basin is still a wheat-productive region today, but is now divided among eight eastern European countries: Hungary, Austria, Serbia, Croatia, Slovenia, Slovakia, Romania, and Ukraine. People brought their plants and animals to the Balkans and then moved up along the Danube River, spreading to the north of the Alps and into eastern Europe. By 4000 BC, farmers had moved from Germany into southern Scandinavia and Poland, and to the British Isles.

Both migratory routes, along the Mediterranean and up the Danube, were already populated with hunters and gatherers. The movement of farmers from the south meant the introduction of new languages and new ways of living, and the immigrants were probably received with the same mix of welcome and hostility that immigrants today experience. The transition to an agricultural lifestyle in the Levant was, at least in some areas like Abu Hureyra, a gradual shift from a nonmigratory hunter-gatherer lifestyle to a combination of farming and hunting and gathering. In northern Europe, the migration of farmers resulted in a rapid and revolutionary change to a new lifestyle. While the hunters in these forests likely also gathered plants, there is little evidence of plant remains in the fossil record. An agricultural lifestyle meant cutting forests and introducing new animals; cattle, goats, sheep, and pigs weren't similar to the red deer, boars, and badgers native to the forests of northern Europe.

In migrating from the Levant to the east and northeast, people journeyed as far as the northwest corner of India. Their movement further into India was limited by the tropical climate. Rain during the part of the growing season when the grain is ripening reduces growth and encourages disease, so wheat and other Levantine crops weren't easy to grow there. Migration was also limited north of the Black Sea. Early farmers traveled with their wheat seeds a little ways into the steppe grasslands of Central Asia, but there the limitations were the high mountain terrain and climate. These grasslands were dominated by nomadic herders, with few traces of agriculture. But some of the herders raised crops in seasonal campsites in the desert or the high mountains, and were possibly the source of wheat introduction to China by around 3000 BC.

The flow of people into new regions had major genetic implications for both our food plants and for human populations. Before agriculture, for the people living in scattered villages throughout the Levant and for small hunter-gatherer bands, genetic exchange had occurred between villages and hunting bands on a fairly local scale. The migration from Anatolia north into the Balkans resulted in a mixing of genetic variation between the immigrants and the peoples living to the north.[2] While the genetic influx from migrating humans affected the northern populations, migration had a different effect on the agricultural species that traveled with humans. Einkorn and emmer wheat were moved to areas outside where they naturally grew; migration with their human partners expanded their geographic range. Plants grow where conditions are right. Plants need sun and CO_2 for photosynthesis, so that's pretty easy, and they need nutrients, which can be found anywhere there's soil. What limits where plants grow is often rainfall or the temperature regime—some species can tolerate cold weather, and others can't.

When seeds are carried to new habitats, they either adapt

or die. Moving from the Levant to Anatolia or to the Euphrates Delta or along the Nile isn't a big change. But moving far north is. Emmer and einkorn are annual plants, so they finish their life cycle in a single year. In the foothills of the Levant, wheat seeds germinate in the fall and grow slowly over the winter, waiting for springtime warmth. In the early days of the summer, the wheats send up a flowering stalk, produce seeds, and die back before the summer heat. The climates of central and northern Europe aren't like the Levant. Winters aren't just cool and moist but cold, and the rains may not start until the summer. The strategy of germinating in the fall doesn't work if plants are killed by cold temperatures over the winter. Moving to a northern environment changed the shape of the evolutionary sieve.

Genetic adaptations occur when a variant in the population has what is needed to survive and thrive in a new environment. With migrating humans carrying pouches of seeds to new environments, another evolutionary force is at play: genetic drift, which is the change in populations caused simply by chance events. The supply of wheat seeds carried to the north is a sampling of the whole population, and some variants may, by chance, not be included. It's an evolutionary sieve of sorts, but one based on chance (which seeds got scooped up and put into the pouch) rather than the result of specific traits that respond to environmental pressures. With the migration, the only wheat that survived in the north was that which adapts to colder climates. Each new climate and set of soils selected for certain traits in the crop species that were unique to that area. This was the beginning of an eclectic set of populations of wheat, each adapted to its local climate, and each contributing to the breadth of genetic variation in the species.

The two species of wheat that traveled from the Levant were einkorn and emmer. These are only rarely grown now, either as a specialty food crop—the ancient grains of the wheat family—or as forage for range animals. Einkorn is the oldest and the simplest of the wheats in this story. It's named for its reticence to procreate, with just a single grain of wheat developing from each little flower (*ein* = one and *korn* = grain, in German). Einkorn grew naturally in the northern part of the Levant, even in the higher plains of Anatolia. When people migrated from the Levant, einkorn went with them in every direction except to North Africa. Einkorn is a shorter-stature wheat and grows well on poor soils, one of the reasons it's grown now for forage. But it can't tolerate hot weather, so the wheats of ancient Egypt that were entombed with mummies for an afterlife snack were never einkorn.

Emmer, product of a hybridization between one of einkorn's close relatives (Latin name: *Triticum urartu*) and a weedy goatgrass, is more complex. Hybrids between species are relatively rare in nature, because most interspecific hybrids are sterile, owing to problems during the formation of their sex cells, the pollen and ovules. To keep the chromosome count from doubling from one generation to the next, all organisms halve the number of chromosomes during the formation of the sex cells. To do this, the chromosomes line up in pairs with a matching chromosome. Those pairs result from having both parents contributing a single copy of each gene. When a sex cell splits into two daughter cells, one of each paired chromosome goes into each of the sex cells, helping ensure that one copy of all the genes necessary for the next generation is present. In a hybrid, the chromosomes can't find a matching partner, because the parents of two different species don't have matching chromosomes. Consequently, the chromosomes don't line up in pairs, and the pollen and ovules get formed with an incomplete set of chromosomes and typically don't function. And even if they do, the resulting fertilized embryo doesn't have all the directions necessary for development.

When *Triticum urartu* got fertilized by a goatgrass pollen, the seven chromosomes of the *T. urartu* ovule formed an embryo with the seven chromosomes of the goatgrass pollen. The seeds matured, and when the autumn rains began, they sprouted and grew into a plant that was bigger and hardier than its parents. What presumably happened next was a mistake that led to a productive outcome. The new hybrid, in the process of forming the pollen and ovule cells, skipped the step of reducing the chromosome numbers by half. Pollen cells had all 14 chromosomes, and so did the ovules. The fertilized embryo then contained 28 chromosomes, twice the number of either of the parent species. And when that plant matured and began to form pollen and ovules, the 28 chromosomes had no problem finding their partner. Emmer wheat was a viable new species, with twice as many chromosomes as either of its parent plants.

Many of the grasses, including the wheat species, can produce seeds by self-fertilizing, and the new hybrid used its own pollen to fertilize its ovules, producing viable seeds. This is the birth story of emmer wheat, a hybridization that took place around half a million years ago.[3] Wild emmer overlapped with einkorn in the southern part of einkorn's range, but grew further south into present-day Lebanon, Jordan, and Israel. Where emmer and einkorn overlapped, they often grew together in large stands with a mix of other grass species.

Doubling the number of chromosomes by genome duplication is a way to rapidly generate a new species. The genome is all the genetic material of an organism; if duplicating a single gene presents an opportunity for new copies of a protein, duplicating the entire genome is that much more powerful for generating material which can be modified in the future. Although it's a rare event in our way of experiencing time, it has occurred often throughout evolutionary history. Two of the major steps of plant evolution from single-celled algae to wheat—the appearance of the first seed plants and the development of flowers—were both preceded by a genome duplication. Genome duplication increases a species' evolutionary potential by providing a vast amount of raw material in the form of organized DNA that's not really necessary. The first set of genes codes for a functional organism; the second set can be ignored, modified, or used in combination with parts of the original set. By having an extra copy of all the DNA, some genes produce additional proteins that can change the development of an organism.

Emmer wheat has an interesting origin story, but nowadays it is overshadowed by two close relatives, durum wheat and bread wheat. Genetically, durum and emmer are similar, classified as the same species, with durum being a subspecies of emmer. Durum, in ancient times, gained a popularity that's lacking in emmer, because of a couple of important differences. While the seeds of emmer are covered in a hard hull, durum seeds have a softer hull that easily falls off, making the seeds much easier to use. We're not certain where durum originated, other than somewhere in the eastern Mediterranean region; but since its seeds needed no charring and pounding to remove the hull, it was quickly favored over emmer. Nowadays, durum wheat is grown for pasta and couscous, which accounts for between 5 and 8% of modern-day wheat production.

But the single most popular member of the wheat family is bread wheat, which first appeared after our ancestors started cultivating the other wheats. Like emmer, bread wheat is a hybrid formed without any help from humans. This time the hybridization happened in a field of emmer growing in the Levant. One of the emmer wheat plants accepted the pollen of yet another weedy species of goatgrass. This time, the 28-chromosome emmer mated with a 14-chromosome goatgrass, another terrible match if you're interested in pairing up chromosomes to form the sex cells. And as with the emmer hybridization, the bread wheat hybrid made pollen and ovules without halving the chromosome numbers. Fourteen chromosomes from goatgrass plus 28 chromosomes from emmer equals a new hybrid with 42 chromosomes. Because bread wheat originated in a field of emmer and was immediately used by early farmers, there are no wild versions of it.

One part of the origin story for bread wheat that is still debated is whether it formed with the hard hulls of emmer, a version of bread wheat that we know today as spelt, or whether the second goatgrass parent conferred a soft hull trait, like we find in bread wheat. If the second story is true, then spelt wheat is the result of a change in bread wheat that caused a reversion to hard hulls. But either way, the loss of a hard hull that stays attached to the seed was a big advantage for people who were pounding and grinding grass seeds for their meals each day, and assured that bread wheat would be rapidly adopted. That is, if it grows well. Bread wheat was harder to grow in less fertile soils and more variable climates. Something that hasn't changed is that we require from our wheat that the plants grow well in the field; that they are as easy as possible to clean and grind; and that the grains, whether used for flour or eaten as full berries, bring qualities to the food that we appreciate.

When people started growing their own food, villages grew larger; more people can live off an acre of farmland than the same area of land used for hunting and gathering. But exactly how they farmed and what they ate is mostly a matter of speculation, at least until the advent of writing. And even once written records became available, those are representative of the large civilizations, organized around agricultural production, and not the majority of the human population living outside the reach of the first city-states, republics, and empires. With that caveat, the first documentation of daily life comes from about 3000 BC, from the Sumerians. The people of Sumer settled in large cities on the floodplains of the Tigris and Euphrates Rivers, an area in southern Iraq historically known as Mesopotamia. Clay tablets inscribed in cuneiform detail farming methods, recipes, notes, accounts, just about everything you might jot down on a scrap of paper. Managing the food for large urban areas required a level of organization to ensure a steady production each year, along with a capacity to store extra grain for the inevitable bad years. Engraved in the cuneiform tablets are directions for planting a field, down to the optimal spacing between rows of grains, the number of seeds per row, and the depth of planting. Oxen were released on the field after spring flooding to trample the soil surface. Workers or slaves then broke up the clumps of soil and smoothed the surface to prepare it for planting. Fields were plowed with an ard, a large hook attached to a wooden beam that was yoked to a pair of oxen or donkeys. An ard doesn't turn over the soil like a plow, but breaks up the soil to create shallow furrows. Sumerians adapted this early field implement by attaching a funnel to the top of the ard handle, so that seeds were simultaneously dropped into the furrow. The first ards were made of wood or stone, and once metals became available, later versions were strengthened by adding bronze or steel to the blade.

Sumerian agriculture is famous for its use of levees and large irrigation canals to manage the flow of the rivers. Mesopotamia received less rainfall than the areas upstream where agriculture first started, so its people corralled the floodwaters in the spring and used that for irrigation later in the season. The Tigris and the Euphrates flowed wild every spring, and the Sumerians imposed a level of control over the force of the flooding rivers with an elaborate system of levees and canals. The levees kept the fields of young wheat protected from spring floods, while the canals brought river water to the fields during the dry summers. The same cuneiform tablet containing planting directions recommends irrigating four times during the growing season: the first after the seedlings emerge in the fall, and the last as the seeds are forming on the plants early the next summer.

Our ancestors' discovery that they could pound and grind grass seeds to make food reset the evolutionary trajectory for a select few species. The wheats, barley, rye, and oats in the Middle East, rice in East Asia, and corn in the Americas have become food staples, collectively responsible for two-thirds of the calories our species consumes. The early Mediterranean civilizations used wheat to make flatbreads, porridges, and beer. In Egypt, bread was made by mixing stone-ground flour with water and a little salt, and cooking it on a stone slab. The staple foods of the Greeks and Romans were barley cakes and a wheat porridge. In Phoenicia, a civilization of independent city-states along the eastern Mediterranean shore around the same time as the Greek civilization, wheat and barley were imported from Egypt. Wheat was made into a porridge that was eaten as a main meal, the porridge made by adding barley, cheese, honey, and eggs to the boiled wheat.[4] And lest it be overlooked, the early ale recipes used lightly baked barley or wheat cakes soaked in water and given time to ferment, creating beverages that likely had an alcohol content lower than our modern beers. Besides being important for feasts and large gatherings, ale and wine had the advantage of being free of the possible taint of contaminated water.

After an initial 40 million years of evolution for the grass family, a few of the grass species got caught up within the labyrinthian sprawl of human cultural evolution. These grasses witnessed the development of small villages that over time grew into urban centers. They were present at the transition from small patches of farmland tended by families to sprawling fields worked by slaves and servants. They were there during the slow development of agricultural technologies, from simple wooden tools to metal-bladed plows that loosened the soils where their seeds were carefully planted.

Emmer and einkorn wheat were moved out of their native range to parts of the world that were hotter, drier, colder, or wetter, selecting for varieties that could prosper in those climates. Some of those wheats still grew wild, free of close human management, and can be found today on dry hillsides of the Middle East. The grasses our ancestors grew to make porridges and flatbreads and fermented drinks became so much a part of their cultures that they focused their creativity on developing better technologies, and their cruelty on increasing the labor force to grow more and more of these grains. With increasing food resources, our populations grew, spreading out across the continents, requiring more fields to be cleared, more seeds to be planted, and more people to work the fields.

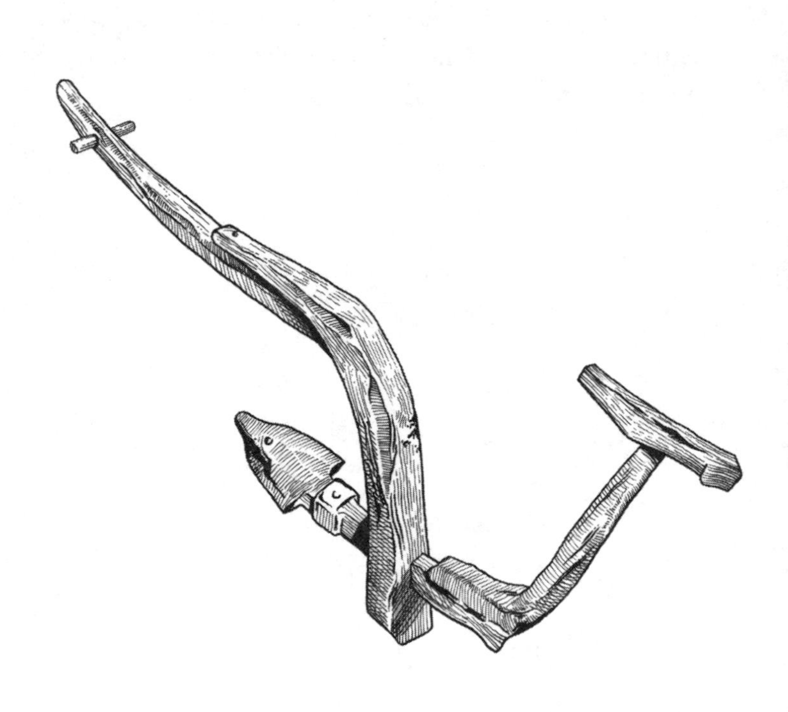

CHAPTER FOUR

From Villages to Cities

Wheat, like all the other grasses and all other plants, evolved in response to environmental stresses. Whether it's a hot and dry summer or a record-breaking cold spring, plants forage for nutrients from the soils, defend themselves from insects and pathogens, and focus what's left of their reserves on making seeds for the next generation. For wheat and other plants that live for only a single growing season, their success is based on their ability to make a good seed. Longer-lived plants that overwinter and grow back from their rootstocks the following spring have a built-in safety net. If the rainfall is too low or the spring temperatures too cool, the plant channels the sugar it builds during photosynthesis into reserves in the roots. It can wait until the following year to make seeds. But the annual plants—most of the plants we grow for food—have no options. They either produce seeds and are present the next year, or they disappear from that site. It's part of what makes the short-lived grasses a better food source; they must produce seeds each year. Rather than investing in long-term storage and deep root systems, annual plants sink all their reserves into their seeds.

When our species started harvesting grass seeds for food, we presented a novel layer of selection. We wanted bigger seeds for more food, and thinner protective layers on the seeds to make the edible parts easier to access. And we wanted wheat in the Balkans, in North Africa, on the high plains of East Asia, areas with different climates from where wheat grew in the wild. We wanted wheat in the savannas, on the fertile soils of river floodplains, in the volcanic soils of the Greek islands, and in the root-bound soils of cleared forests near the Danube. Each environment was a new evolutionary sieve, adding a swerve to wheat's evolutionary trajectory. The selection pressures were different in the Balkans than they were along the Nile. Exposing a single plant species to many different environments is a perfect way to generate, intentionally or not, different varieties of wheat. And the genetic makeup of emmer and durum, and even more so of bread wheat, was excellent for generating new varieties. These wheats were formed with the chromosomes from two or three species, resulting in more genetic material for generating variants that can grow bigger, forage deeper, and evade predators.

What we know of the genealogy of bread wheat and emmer and durum is thanks to our understanding of genetics and chromosomes and mating systems. With science as our backdrop, we can piece together the story of how a pollen grain from a species of goatgrass fertilized an individual of emmer wheat in a Levantine farmer's field to form bread wheat. Next, with a theoretical and empirical understanding of polyploidy—how organisms increase their chromosome number—we can fit that story into our understanding of species barriers and adaptation to changing environments. And so we develop a vocabulary of technical terms to describe these patterns of plant variation, and the mechanisms for how those patterns are maintained. The wheats that were grown in the Balkans, in Egypt, and in France were landraces, differentiated from each other as a result of growing in a particular area where farmers and the environment functioned as a selective force for specific traits. Farmers had understood the importance of landraces thousands of years before we had the capacity to look inside cells under the lights and lenses of a microscope to count their chromosomes. Theophrastus, a pupil of Aristotle and a prolific scholar, cautioned during the third century BC that seeds from plants grown in lands with a colder winter will be slow to produce seed; hence the crop would likely be destroyed by summer droughts before the grain was produced. "They are hardly suited to local conditions in respect to time of sowing or of germination," he advised.[1]

When our ancestors began growing wheat, the grass family had already had 40 million years to perfect its gig. If a species couldn't manage the challenges of rocky soils, low rainfall, and hungry animals grazing on both leaves and roots, it wouldn't survive. But the relationship with humans represented a new challenge. Besides moving wheat to new habitats, our ancestors wanted to grow a lot of it, all in one spot. To that end, they cleared fields, added nutrients and water when needed, and then sat back to enjoy a big harvest. Well, not exactly sat back. Farming, particularly pre–John Deere farming, is hard physical labor. It is in fact what causes many anthropologists to wonder why Neolithic peoples started farming, given how much work was involved in a world where the only tools were what they could fashion out of rocks or wood.

To facilitate the production of a lot of wheat in one spot, the first step is to clear the soil. If you've ever tried to dig up some of your backyard to make a garden plot, you'll understand why your neighbors rented a rototiller for making theirs. It's a slow process, breaking through the roots to loosen soils and create a welcoming growing medium. And that's with a steel-bladed shovel. In contrast, the first tools at Abu Hureyra were likely wooden digging sticks, used to create tiny pits or furrows for the seeds. Stone-bladed hoes were depicted in artwork by the Sumerians and the Egyptians. These handheld hoes were eventually modified to make the ard, an early version of the plow. The elongated arc of a blade was pulled through the soil, by humans or by humans guiding livestock, depending on their wealth. Ards were first made of branching hardwoods chosen for durability and easily replaced if they broke. They were good for creating a shallow furrow through the soil and are still in use today in parts of the Mediterranean region, particularly in light soils that could be damaged by a steel-bladed plow. With developing knowledge of how to smelt and cast copper, and later bronze and then iron, the blades of hoes and ards were sometimes constructed or lined with metals to increase their strength. In fields that needed to be cleared, possibly after a year of rest from growing crops, the ard was pulled through the soil first in one direction and then crossways across the field, with a couple people following behind to further break up clumps with hoes or, in hard soils, large hammers.

The easiest soils to clear were along the floodplains of the major rivers, where each year rushing waters scoured the plants and lay down new sediments. Consequently, the earliest large-scale agriculture took place in the fertile soils along the Tigris and Euphrates, the Nile, the Yellow, and the Indus Rivers. Alternatively, farmers clearing fields in forested areas needed to wield stone axes, then burn the branches and small trees to clear the debris and break down the thickest trunks. With the exception of the tropics, cleared forested soils have a stock of nutrients that can support a healthy growth of crops. But the first years after clearing the soils are still what the ancients described as heavy; only plants with strong root systems could grow there. Emmer and einkorn fared better on denser soils than bread wheat.

Once the field was cleared and the soil broken up, seeds were either scattered across the surface or planted into furrows. To protect them from hungry birds or a drying sun, farmers covered the seeds with a light layer of soil, using either another pass of the ard, willow branches dragged over the fields, or a plank or branch with tines added to comb the surface. Another option was to let oxen trod the fields, pressing the sown seed into the soil.

Plants need water to grow. The first crop plants of the Levant evolved in dry climates. Wild wheat and barley germinate in the fall, grow slowly over the cool and moist winter, and make their seeds and die before the summer heat and drought. Lentils and chickpeas need a little more rainfall than do the cereal grains, but nonetheless grew in the wild in very dry climates. The problem for people depending on farming for food was the inevitable dry years, especially in parts of the Levant where rainfall was less predictable. Even the earliest agricultural villages have evidence of ditches that could have been used to drain or bring water to fields. Along the Tigris and Euphrates, the annual rainfall was lower and less predictable, leading Sumerians to build walls to create basins for capturing floodwaters for use later in the growing season, and to protect fields from damage by rushing floodwaters.

Irrigation along the Nile River extended the area of farmable lands, bringing water to desert fields. The Nile is over 4,000 miles long, and picks up water from 10 African countries before it reaches Egypt and flows the final 1,200 miles across the Sahara to the Mediterranean Sea. The rains that fell in the highlands of Rwanda and Burundi reached Egypt by midsummer. In fact, the waters that overflowed the Nile's banks, carrying fertile soils from upstream, were so regular that the ancient Egyptian calendar was divided into three periods: the period of flooding, the period of crop growing, and the dry period for harvesting crops. Irrigation along the river served dual purposes: the floodwaters could be retained behind walls until just before planting, maximizing moisture for the crops; and by constructing a series of basins along a slight downhill gradient moving away from the river, the floodplain was extended into the dry zone, increasing the area for growing food. However, one advantage of the floodwaters—fresh sediments and nutrients laid down each year—was equally a disadvantage for managing irrigation structures. The annual sediment delivery raised the height of the basin floor incrementally each year; eventually, the basin floor was above the level of the riverbank, rendering the basins useful only during extremely high flow years. The solution was to connect each basin to a water channel upstream at a higher elevation, requiring coordination from people at different points along the river. On a local scale, that meant cooperating with the neighboring village. But on a larger scale, how floodwaters were managed hundreds of miles upstream had a direct impact on all the villages downstream.

Wheat, like every other plant, is made up mostly of water. If you get rid of that water by drying plants in a low-temperature oven, what remains is mostly carbon, oxygen, and hydrogen. Beyond that, a little bit of what is left, around 2%, is nitrogen and, in decreasing order, potassium, calcium, magnesium, phosphorous, and sulfur, and so on in incrementally trace amounts. Carbon, oxygen, and hydrogen are easy for a growing plant to access from the atmosphere and from water. Nitrogen and other elements all come from the soils. The reason we eat seeds is that they're packed with nutrition, enough resources for the next generation to put down its first roots and unfold its first leaves. After that, the seedling is on its own to gather what it needs. The roots have their job cut out for them in that regard, with some crucial help coming from the hardworking, photosynthesizing leaves.

The carbon dioxide (CO_2) for photosynthesis enters the leaf through donut-shaped openings on its surface. Those openings are carefully regulated by the plant, because water evaporates from the plant while CO_2 is diffusing into it, a life-and-death consideration in sunny, dry environments. If there's enough moisture in the soil, the water evaporating from the leaf is an indispensable benefit to the plant. The loss of water to the atmosphere creates a negative tension inside the leaves, like the negative tension that's created when you drink water through a straw. From the tips of the leaves to the depths of the roots, narrow columns of water are supported inside the long xylem tubes. When water evaporates from the leaves, the water columns are pulled up through the xylem tubes. The negative tension is felt all the way down to the roots. The resulting shortage of water in the roots is why soil water moves toward them.

Water evaporating from the leaves is linked to plant nutrition because nitrogen and most other nutrients are dissolved in the soil water. When evaporation creates a negative tension to pull water through the plant and attract water to the roots, it also brings with it a nutrient solution. The root cell membranes then decide what to take up and what to exclude. If the soil is dry and a plant becomes water stressed, it can close the leaf openings to conserve water. That also reduces the plant's access to CO_2, and hence its potential to grow. The search for water and nutrients is what keeps the plant extending its root system deeper and wider through the soils.

Assuming there's adequate water for plant growth, soil fertility is the next consideration for growing a lot of wheat in one place. With each successive harvest, we take away some of the nutrients from the soil for our own bodies. With sunlight and rain and the help of organic acids released by plants, soils slowly release minerals, but not at a pace that keeps up with what we remove with each harvest. How the first farmers managed the soils to replenish the nutrients, we will never know. Perhaps they initially farmed their fields until their crops didn't grow well, and then cleared a new field. It's a variation of slash-and-burn agriculture, and fertile soils likely could support crops for several decades before the yields declined to the point that the work it took to farm there was no longer worth it. The archaeological record shows that villages were sometimes abandoned and then reoccupied centuries later. Soils depleted of nutrients can be restored by growing plants that slowly build and return those nutrients until soils are fertile enough to support farming. In hot and dry regions, the recovery can take decades or even centuries.

Short of abandoning farmland altogether, the practice of leaving fields uncultivated for a year or two is known to have been in effect at least since Greco-Roman writers recorded their agricultural methods. Such fields fill with whatever plant species can get there first, usually weeds. After letting their fields rest, farmers prepared the soil for replanting crops by passing an ard through it a few times, or they let their livestock graze the weeds before they plowed any plant remains back into the soil. Dead plant material becomes soil organic matter, which soil animals and microbes break down to release its nutrients back into the soil. But even agricultural fields are subject to the law of conservation of mass. Other than carbon, the nutrients released in decomposing plant material were originally obtained by the plants from the soil. The same is true of the manure left by grazing animals. Manure is a good source of fertilizer for the fields, but its nutrients are what's left from the original plants after the grazers' gut system extracted what these animals needed to support their own muscle mass. Although plants can return organic matter to the soils, at best they can just cycle the nutrients already in the system.

The exception, the rather miraculous exception, lies within a group of plants that form a symbiosis with specialized bacteria, the nitrogen fixers. Nitrogen surrounds us, comprising almost 80% of the air we breathe. Plants quite handily extract CO_2 from the atmosphere, so why not nitrogen? The difficulty is that as a gas, nitrogen is tightly bound to a second nitrogen atom by three shared electrons that orbit the pair of atoms. Plants and animals use nitrogen in forming the amino acids that make up proteins. Proteins orchestrate much of what happens in our cells. But there is no plant or animal that can break the triple bond of gaseous nitrogen to make it available to form amino acids. The only organisms that can access that nitrogen are nitrogen-fixing bacteria that contain an enzyme, nitrogenase, that splits the triple bond and converts the nitrogen into a usable form.

Lentils and peas have managed a way to attract one of the few species of bacteria that can use atmospheric nitrogen. Peas and beans and other plants in the legume family create a small home for those microbes in specially constructed nodules on their roots. For the bacteria's nitrogen-fixing enzyme to work efficiently, they need a low-oxygen environment, hence the special chamber, which is lined with a hemoglobin-type molecule similar to the one that carries oxygen in our blood. Inside the nodules, the plant hemoglobin picks up the oxygen to take it out of circulation. Splitting the triple-bonded nitrogen gas is enabled by nitrogenase, but it still requires a lot of chemical energy. Single-celled bacteria don't have big energy reserves; plants, however, with their unique capacity to tap into sunshine, can easily supply the needed energy in the form of the storage molecule ATP, generated during photosynthesis. Microbes access nitrogen thanks to their unique enzyme, nitrogenase, and a symbiosis with a plant that supplies the energy.[2]

One reason that agriculture-based civilization in the Middle East prospered and spread in all directions was the combination of wild plants our ancestors domesticated: grasses with easily stored seeds and nitrogen-fixing peas and lentils. Growing legumes along with wheat reintroduces some of the nutrients into the soils, keeping the fields usable through more harvests. People relying on their small fields for a major source of food would notice when seed production was higher or lower. That wheat plants grew better in a field where peas or lentils had grown the year before would be enough of an impetus to start rotating crops, with nitrogen-fixing plants one year and grains the next—a practice that continues to this day.

Once the plants have enough water and nutrients to grow, the next challenge is harvesting large quantities of seeds in one place. The seeds on wheat and other grasses are clustered at the top of the plant, in what's called the seed head. Wild grasses drop their seeds naturally, so early seed collection, before einkorn or emmer were domesticated, was likely done by beating the seed heads over a basket in the field. After domestication, seeds remained attached to the plant, so early farmers harvested wheat by either pulling the whole plant out of the ground; stripping the seeds off the plant in the field; or cutting the grass stalk and bundling the seed heads and stalks. The earliest tool for cutting grasses was the sickle, a handheld tool with a curved blade sharpened along the inside of the curve. The earliest sickles were made from antlers or hardwood, and some toolmakers improved the blade by carving a groove on the inside of the curve, then inserting pieces of sharpened flint held in place with soft asphalt. Wheat could be cut near the seed head, since the seeds were the important food source, or near the ground, as the stalks were used for bedding, fodder, baskets, and roof thatching. If just the seeds were harvested in the field, the stalks could be cut later for other purposes.

Harvesting crops, as with sowing seeds, is done with an eye to the heavens to keep track of the weather. While seeds need to be planted when the soils are moist but not too wet to work, wheat must be harvested when the weather is dry, but not too dry. These requirements necessitate efficient harvest methods, as each day has the potential for crop-ruining rainfall or hail. Thus, extra hands were needed during seeding and harvesting times.

Besides the vagaries of bad weather, harvested grains can be destroyed by small mammals, birds, and insects, the latter often carrying fungus that can also extract energy from the seeds. Therefore, processing the grain and putting it into safe storage was a priority. With the seed heads in hand, the next step is threshing, or beating, to detach seeds from the stem of the seed head. Threshing floors were smooth, tiled areas where the wheat stalks were laid down and either beaten or trampled. Oxen or mules or horses walked over the stalks to break them up, or sometimes a sledge was pulled behind the animals. *Sledge* is not short for *sledge hammer*—it's like a sled or a chariot built of wood, about 2 yards long, with a flat surface that's dragged over the layers of harvested wheat. To increase the sledge's efficiency, our ancestors embedded flint teeth along its underside, and someone sat or rode on it to increase its weight, the better to break apart the stems and free the grains. Much later, they developed the flail, a handheld rod with an attached swinging arm used during the Middle Ages to pound the stalks.

When the threshing was finished, the broken mess had to be separated into that which could be eaten and the rest, which we call the chaff. For that, the location of the threshing floor was important. At the top of a hill or in an area with a dependable breeze, air currents could sort a pile of the broken straws that had been tossed upward into the lighter chaff and the heavier grains. What wasn't in the "save" pile could be fed to the animals or used in other ways—to temper plaster or clay bricks, to line the storage pits. That process of winnowing was followed by more sorting and cleaning by hand and with sieves to remove weed seeds, stones, and pieces of hard stem.

Grass seeds are a good food source because they can be stored and then ground when needed. Their high-protein embryo and high-starch storage tissues also make them a good food source for other organisms. But the farmers' intention with the harvest was not to share what they reaped with other animals or microbes. So our ancestors stored their wheat seeds in baskets, in containers made of clay or cow dung, and in pits built either into the walls of their mud-brick houses or belowground. Of the different storage methods, belowground pits are most likely to be discovered by archaeologists. These were sometimes lined with some of the cereal remains left on the threshing floor. During the Middle Ages, people burned dried plants or fuel inside the pits to harden the walls and eliminate hungry organisms or molds before filling them with the winter's grain stock.[3]

Grass seeds are small, able to be stored for longer periods than most food, and, compared with other foodstuffs, relatively easy to move. It's no coincidence, then, that our earliest civilizations and cities were based on farming one of the grasses—wheat, barley, millet, rice, and eventually corn. The advent of farming altered the evolutionary trajectory for the wheats and their companion species, but it also allowed for more options in the social trajectory for groups of people. People had lived in villages even before there was agriculture, but farming supported larger populations on the same area of land. Six thousand years after Abu Hureyra was first settled, people were amassed in the first urban areas, cities with 40,000 to 50,000 people. While we take cities for granted now, the reasons for the shift from small settlements to large urban areas are not intuitive.

The earliest cities developed downstream from Abu Hureyra, in an area where the Tigris and the Euphrates Rivers flow toward the Persian Gulf. Both rivers were large; the Tigris flowed deep and fast, and the Euphrates flooded each spring with snowmelt and rainwater from its 2,000-mile journey from the mountains of Turkey through present-day Syria and Iraq. People had been living for thousands of years in the rich and fertile wetlands between the two rivers. Similar to the area around Abu Hureyra, the ecologically rich zones along the rivers provided edible plant roots and seeds, birds and waterfowl, and small mammals and migratory gazelles, along with fish and crustaceans from the rivers themselves. With their agriculture and a diverse and abundant wild food source, people could live well without an overarching political authority. The region lacked a good source of stone and large timbers for building, so settlements were constructed with plant materials gathered along the river, leaving few traces in the long-term record.

The earliest civilizations, including those of Egypt, the Indus Peninsula, and China, arose in river valleys. The first large-scale irrigation canals along the Euphrates for which we have clear evidence date to around 5000 BC, and required an estimated 5,000 hours of labor to construct. It's likely that the canals were built near a fairly large settlement. With irrigation, farmers could produce more than enough food to feed the local people, and that surplus meant two things. Populations along the river increased, as did trading. Wood from Lebanon and copper from areas upstream in modern-day Turkey were transported downriver, and animals and wool were traded with nomads and pastoralists living farther away. Trading also meant that there was a market for artisanal goods, and an opportunity for people to specialize in making things that could be exchanged for food.

There are many ideas about why the first urban areas developed. One possibility is that shifts in the climate resulted in a decline in water levels in the Tigris and the Euphrates, reducing the size of the floodplain and area for growing crops. Without abundant water to flush the fields, salts accumulated there, further reducing people's capacity to grow food. With a food shortage, people are more inclined to accept political authority, and an organized state could provide the infrastructure for urban areas to produce and distribute food. Additionally, an organized military could both protect and coerce the people residing within the state.

The development of a political state is generally accompanied by a stratification of wealth, a centralization of power, a religion or some kind of belief system to justify the centralization of power, and a military. Land records from Sumer dating to around 3000 BC show that tracts of irrigated land were distributed in sizes larger than what a single family could farm. If those tracts were managed by an extended family group, the development of economic prosperity, enabling families with a particularly good allotment to accumulate wealth, was possible. During climate shifts to lower rainfall, farming in that area would be viable only on irrigated lands, further concentrating people near the food sources. Some scholars have argued that states arise from the need for irrigation structures; however, the knowledge necessary to construct these systems had developed before the formation of states, as people dug canals and built walls on a smaller scale as needed. What the state could provide was centralized organization for the labor necessary for constructing and maintaining the large-scale irrigation schemes.

As people concentrated into large urban areas, the result was increasing numbers of hungry stomachs in one place and fewer farmers. But the most basic requirement for political stability is a well-fed population. Enter the temple, which in Mesopotamia served as an urban area's organizing center, equipped with scribes that functioned as irrigation managers and record keepers. Scribes recorded their first accounts by using hard cylinders containing incised symbols. They rolled those cylinders on a clay tablet to create images depicting goods that were exchanged or stored or distributed. Written language developed from there; scribes used stalks of wetland plants to inscribe pictographs on clay tablets, with the symbols becoming increasingly abstract until they finally mimicked sounds, particularly those of the scribes' names. Writing was initially practiced only by the scribes, so the scripts recovered from Mesopotamia are mostly lists—of goods, of properties, and of methods.

The advantage of being part of the state was clear for those who lacked the land or the wherewithal to grow their own food, especially if wild sources for hunting and gathering had disappeared. But the disadvantage for farmers was that they had to grow grain not just for their family; a portion of their harvest was collected as taxes. That portion was often 20%, but in Egypt the amount varied between one-third of the grain produced on small farms and up to one half on the large estates. The state served as an external motivation to move beyond subsistence farming that provides for individual families; with state oversight, each farmer had to produce enough surplus to provide for the rest of the people.

Ownership of the land where foods are grown was as big a question for the Sumerians, the Greeks, and the Romans as it is today. Property distributed by the state to people with power was a way to maintain state authority and at the same time stabilize the economic inequalities between social classes. But then there comes the question of who works the land, breaks up the soils, plants the seeds, weeds the fields, and harvests the golden grains. Labor on large estates and on state lands could be hired, but it was often coerced—extracted as a tax or a punishment. Unpaid labor was required of people who were unable to pay off their debts, debts often caused by a poor harvest year that left the farmer to choose between paying his taxes or feeding his family. The state could also expand its food stores by conquering other villages and their agricultural lands. The conquered people were then forcibly relocated to work the fields within the walled state. This was of course a form of slavery, but in many of the early states, slaves had the possibility of moving out of that role. The need, for example, for a growing military required the use of farm laborers, and as agricultural workers became soldiers, the state resorted to more raids to replenish the labor pool. Just as wheat had become the commodity that could be traded for metals and precious stones, so humans represented a labor commodity, and early civilizations were built on the exploitation of other peoples.

Since the early days of farming, the inequity between the landowners and the laborers has been an obvious point of contention for field workers and peasant farmers. History is replete with stories of peasant farmers losing their land to wealthy landowners. Land reforms and political protections were passed at times throughout Greek and Roman polities, to redistribute lands from wealthy property owners and to protect farmers from enslavement. But attempts to increase political stability through social equality were never long-lived, and reflect patterns in the social history of farming throughout the world. Over the last decades, the story continues to repeat itself. Family farms have been lost in the United States and other parts of the world because of debt and the purchase of land by increasingly larger farming estates.

Besides forcing people to labor in the fields, the role of the military relative to agriculture was to protect food stores. For people who lived outside the early states and weren't inclined to hoe fields and thresh grain, the coffers of grain, whether placed within a temple or in a farmer's storage shed, were a big temptation. Consequently, states needed military power to defend their stores, and they built walls around their cities. Farmers needed protection too, and by paying a tax could be offered some protection from barbarians. In this way, the state extended its influence outside its walls, but only to the areas near a convenient transportation route. The Sumerian cities of Mesopotamia, the Egyptian settlements along the Nile, the Chinese dynasty clustered along the Yellow River, and the Indus civilization in what is now Pakistan arose along rivers. The rivers supplied water, renewed nutrients in the fields, and served as a transportation route for delivering food to a central store and for shipping trade goods.

The development of large cities and a structured society was dependent on rivers for irrigation and transportation, and equally dependent on the seeds of grasses used as food: the wheat and barley of the Levant or the rice grown in China. These seeds were high in protein as well as easily stored and transported, and they created the potential for elaborate exchange networks. Grain collected as a form of taxation could be more easily regulated than other kinds of food for that purpose. Tax collectors could scan a field of wheat early in the season and estimate how much grain would be produced. Taxes were sometimes levied based on the fertility of the farmers' soils, adding an incentive for people to farm the lands to optimize production. To that end, the state, from the time of the Sumerians onward, dictated best practices to maximize the production of wheat and barley to fill the coffers. The quest for practical knowledge of how to best grow food had shifted from the anticipation of hunger in the winter months to the demands of the power holders for a larger surplus for engaging in trade of both necessary and luxury goods.

As wheat became a staple of cuisines around the Mediterranean Sea, it also became a commodity to be traded for other foods. Greece, with its thin layer of low-nutrient soils, had the right climate for growing wine and olives, but depended on trade for a steady supply of wheat. Athens imported wheat from Egypt and areas around the Black Sea. Even Sparta, surrounded by large landholdings, had to import wheat from Sicily and the Black Sea region.

The Romans' appetite was felt across North Africa all the way to northern Europe as the Roman Empire expanded in part to annex lands that could be used for growing wheat and other cereals. Sicily, Sardinia, and North Africa supplied much of the wheat for the empire, and areas southwest of the Rhine and south of the Danube exported wheat, rice, and spices downstream to Roman settlements. Along the Danube, villages also exported food to Roman settlements, while those too far inland to make transport of goods feasible were ignored. The collapse of the Roman Empire freed those outposts to focus on their own cuisine, which by that time had expanded to include the fruits of the Mediterranean region that could grow in the northernmost reaches of the former empire. When Roman rule declined, wheat farming decreased across the European continent. Thereafter, subsistence farmers grew rye and oats more commonly than wheat, since these more reliably produced seeds in the heavier soils of previously forested lands.

The oldest preserved details about farming methods and food preparation are derived from early civilizations who prescribed farming techniques and valued written records. But in reality, most people living around the time of the Sumerians, the Greek city-states, and the Roman Empire lived outside the influence of organized political states. That remained true until about the seventeenth century. Most agriculture was being practiced by subsistence farmers, often with livestock to help with the labor and serve as a backup food if crops failed.

Wheat, and our capacity to gather and plant its seeds, enabled the production of food on a scale large enough to support the development of states and powerful political regimes. Innovative farming systems were developed long before any science of agriculture, as people's comfort and survival depended on the artful management of food production. Early farmers must have understood their crops: noting the differences between emmer or einkorn or bread wheat, knowing when growing rye or oats would be more likely to yield a full coffer for the next winter or dry season, and maximizing crop rotations to also provide feed for the animals that helped with the labor. Early farmers cleared fields and turned over soils to grow grains and pulses, and kept smaller gardens that could be intensively managed to produce a bountiful yield of vegetables and spices that would complement the staples.

Farming methods, then as now, depended on the scale of the farming operation, and small plots could be managed more intensively than large fields. Early farmers grew crops to feed their families, so the years when crops didn't grow well were painful ones. A farmer's awareness that he had harvested more wheat from a newly cleared field would justify his leaving a field unplanted for a year after a crop, letting the wild plants take over, and then plowing the field to add a crop again. The practice of fallow had been common by the time of the Sumerians, and was used by the Greeks and the Romans. It continues to be a common practice in certain parts of the world, including arid regions of the United States. In fact, some of our most innovative organic farming systems heralded in the sustainable farming and healthy eating tomes apply the same practices developed in the fields of the Mediterranean region thousands of years ago. Animals and crops were part of an integrated farming system. Larger animals were an important part of the necessary labor, so the fields that were rested between food crops were often allowed to fill with plants and then were grazed. Manure deposited by the grazers returned some nutrients to the soil.

As with most relationships, our relationship with wheat has grown more complex with time. Wheat has come a long way in our story. Agriculture supported the development of civilizations, and wheat was one of the main crops grown to feed the city dwellers and for use as a commodity to be shipped across large bodies of water and exchanged for other goods. Early civilizations saw the development of writing systems, the first inquiries into science and philosophy, and the exploration of arts and architecture. Wheat, along with barley and peas and lentils, served as the basis for the development of large urban areas on both sides of the Mediterranean Sea. People honored this humble grain; the Egyptians left bowls of emmer wheat in their pharaohs' tombs. Along with grapes and olives, wheat was the basic nourishment for the Greek city-states, a culture built on the primacy of wine, oil, and bread. It was a main staple of the Roman Empire that spread from Rome to North Africa, through the Middle East, up into the Balkans, and throughout continental Europe. Even the settlements not directly under Roman control understood the advantages to growing the crops that could be traded to the Romans and their ever-expanding appetites.

But wheat and humans, despite the cultivation of soil in large fields and the construction of cities, are still subject to the forces of nature. In the natural world, one action has a ripple effect up and down the food chain. Our predilection for flat breads, for gruel, for ale, and for white bread and noodles has transformed the planet in increasingly intensive ways. The urge to feed ourselves and the option to trade food for luxury items have been the motivating forces behind these changes.

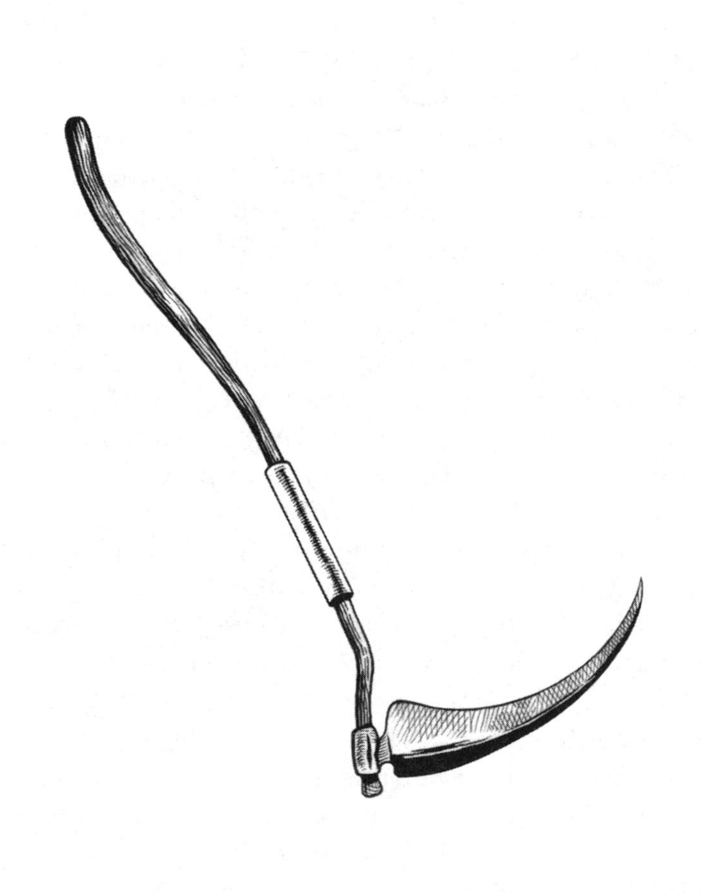

CHAPTER FIVE

Relationships Are Hard Work

The story of wheat has become inextricably intertwined with our human story. The management of food systems, particularly when the networks connecting distant peoples were sporadic or slow, was the difference between societies prospering or failing. And how societies structured their labor systems—in terms of forced or free labor and the attendant working conditions—provides an impression of the quality of life for people living within or on the edge of those societies. People's need for food was the organizing structure of political and social systems that grew up around the birthplace of wheat cultivation in the Middle East and the Mediterranean region.

As the Roman Empire declined over the early centuries of the Common Era, it split into two regions. The eastern portion melded into the Byzantium Empire, centered in Constantinople. The Byzantines were a diverse culture that embraced Greek and Latin scholarship, and continued the Mediterranean farming systems of trees, vines, and cereals, well into the sixteenth century. Wheat continued to be farmed in much the same way as before, in fields worked with ards and in crop rotations with legumes. The incursion of Islamic influence, with the takeover of Syria beginning in the seventh century, introduced variety into the regional cuisine. Arab peoples, living in warm regions closer to the equator, built irrigation systems based on Roman structures, but organized more intensively. Underground canals and water transport methods, including animal-drawn wheels, provided for food year round, and even for multiple harvests in a single year from the same field. Arab peoples contributed new foods to the Mediterranean cuisine, introducing plants from Africa, including sugarcane, eggplant, spinach, artichoke, and watermelon. And most important for our story, they brought durum wheat, a variation on emmer but without the hull.[1]

For bread wheat, it was the western half of the fallen Roman Empire that took the grain on a new trajectory, and where our story continues. The political context for this part of the story begins with a disorienting series of wars and takeovers, atrocious slaughters and pointless loss of life. The eventual result was that the bulk of the European continent came under the control of Charlemagne, who was first King of the Francs (people from an area that included the land that is now France and Germany), then King of the Lombards (adding on northern Italy), and in the last 14 years of his life was crowned the head of the Holy Roman Empire (tack on modern-day Austria and the Catalonian portion of Spain).[2] Charlemagne's rule marked the beginning of the feudal land system in Europe, where kings allotted land to trusted nobility in exchange for military support. The lords then divided their allotment among key military people, the vassals. And like a set of Russian nesting dolls, the vassals divided their lands among the peasants. The peasants lived on the land sometimes as freeholders and other times as serfs, who were bound to working on the property and subject to its owner.

Beginning with the Sumerians and through the Greek and Roman times, farming wheat in the Mediterranean region was dependent on a mix of free and forced labor in the fields. During the medieval period in Europe, whether these workers were free or bound varied with the practices and philosophy of the particular lord and vassal who oversaw operations on the land. Serfs were technically not slaves, but they were also not free. During the early medieval period, the conditions for serfs in eastern Europe were better than for serfs in the west, primarily because eastern Europe was only sparsely settled. To attract peasants, the lords had to offer good terms. But working conditions were to change. By the sixteenth century, eastern Europe was notorious for its poor treatment of serfs, who had lost most of their rights and could be sold by their owner or shipped to Siberia at the lord's discretion. As far as having slaves work the fields, Christian law throughout the European continent declared that Christians could not be enslaved, a step in the right direction. But that left room for the enslavement of Muslims and Jews, along with Christians who either had committed some wrong or were adherents of some outsider variant of Christianity, such as Orthodox Christians from eastern European countries. It took the ideas from the eighteenth-century Enlightenment and accompanying economic rearrangements before slavery and serfdom were eliminated across Europe in the nineteenth century.[3]

The feudal system was a form of barter—the right to land in exchange for military service. Warriors were a key part of the medieval social structure, necessary for a couple of regular activities: to defend the lord's territory and to acquire the adjacent territories. The entire medieval period was violent, but the tenth century in particular featured raids by Vikings from the north, Arabs from Spain and the southern boundaries, and Magyars coming from the east. Soldiers required horses and armor, and armor and other metal goods weren't cheap. The cost of a coat of armor in the eighth century, for example, was equivalent to that of 15 horses or 23 oxen. To support the armed forces, taxes were levied on the peasants and serfs. These might include a toll for the birth of a child, the death of an animal, the marriage of a daughter, inheritance, using the lord's mill to grind grains into flour (milling elsewhere was sometimes prohibited), and on and on.

Serfs were required to give a portion of their crops to the lord—not so different from the taxes levied on them—but they were also obliged to work at the lord's request, and were often bound to a specific piece of land, which limited their right to travel. Compulsory labor could include a specified number of days of plowing and sowing, or tasks like gathering wood for the lord and fencing his gardens. It was not against the law for lords or vassals to use violence on serfs who neglected to pay their taxes or meet their work obligations. At the same time, the landlords were obliged to protect their tenants from those who would potentially inflict even worse conditions. It wasn't an easy life, but conditions for serfs improved slowly across the centuries of the medieval period, particularly when labor shortages occurred, compelling lords to treat serfs more humanely.

While farming in the Mediterranean region relied on the ard and oxen, the ard was largely ineffective in the forested, wetter soils common throughout northern and central Europe. The greatest change for the wheat fields moving north from the Mediterranean was the heavy moldboard plow. This plow differed from the ard in having two blades: the first cut a vertical slice through the soil; the second, set at an angle closer to horizontal, sliced through roots and soil. The wedge of soil carved by the two blades pushed up against an angled backboard that flipped the soil over, creating a ridge of turned-over soil and a furrow from which the soil was cut. This microtopography was beneficial in soils of northern Europe that had a higher clay content and, with that, an often greater moisture content. The ridges raised the soils so that water could more easily drain away, and the furrows sheltered seedlings from winds and during dry seasons reduced the evaporation of existing soil water.

The moldboard plow was used on the land managed by Charlemagne in the ninth century, but its introduction in other parts of Europe was slow, taking centuries to spread across the continent. The adoption of this plow was partly limited by the cost and availability of iron. Iron had been mined and forged in Europe for over 1,500 years by the time of Charlemagne's rule, but was used to make armor and weapons for the well-to-do. Reallocating precious materials for food production instead of arms production wasn't a fast or easy change when there were serfs who could provide the agricultural labor for practically nothing.

Another farming approach advocated during Charlemagne's reign that eventually spread across Europe (though at a rate not much faster than that of the heavy plow) was a three-year field rotation. Instead of planting the same crop year after year or alternating a year of wheat with a year of lentils, another year of fallow was added to the rotation, in which the fields were allowed to fill with whatever plants colonized, and then were used as pastureland. When fields were left fallow in Mediterranean regions by the Romans, the Greeks before them, and the Sumerians before them, it was in part to store water. In northern Europe and other regions that had regular rainfall, a year of pasture added organic matter to the soil. What this looked like across the landscape was that one-third of the fields would be planted with wheat, another third with lentils, and the last third managed as pastureland.

The improved farming methods and the increase in acres plowed into farmland went hand in hand with an increase in the population of Europe; in the three and a half centuries between 950 and 1300, the number of its inhabitants almost tripled. All those people required more food, and at the same time provided more labor to work in the fields. In addition, the economy was bolstered by the changes in agricultural technology. The iron plow; the harnesses for teams of animals to pull the plow; the blades for scythes used to harvest hay and for sickles to harvest grains; and the horseshoes and harrows all created a demand for artisanal expertise to forge the metals and design the tools. Villages grew, housing people who made their living not from farming the fields but from making the tools and weaving the cloth, and serving as a marketplace for farmers who increasingly were paying their rents and taxes in cash instead of in produce or grain.

Regardless of the improved ways to work the soil and the 3-year field rotation method, every ecological system, whether it's natural or managed by humans, is subject to the same laws of the universe. Removing nutrients from farmland without fertilizer input can go on for only so many years before the soil nutrient levels drop low enough to affect crop yields. Aristotle (incorrectly) understood that plants absorbed the organic matter in the soil, a conclusion based on observations that plants grew better in soils containing organic matter. Early farmers around the Mediterranean Sea grew peas and lentils, crops that fix nitrogen, and they kept their fields fallow to retain soil moisture, reducing the rate at which nutrients are removed; they also added manure to fields. A tenth century Byzantine farming manual, which included a collection of Greek writings, recommended the use of manure, specifically after the manure was composted with ash from the hearth, chaff from the winnowing of crops, and dung from as many animals as could be collected, including humans.[4] In northern climates, animals were kept in stables during colder weather, and were fed hay that had been harvested the previous summer and fall. With penned animals, the manure from the stables was easily collected and composted before spreading on the fields.

The growing population of Europe in the eleventh, twelfth, and thirteenth centuries and the attendant rising quantities of wheat and other foods harvested each year were partly due to the technological breakthrough of the plow, the innovation in crop rotations, and the application of fertilizer in the form of manure. But there's one other important factor. The climate across continental Europe was warmer and drier during this time, a period known as the Medieval Warm Period. This period translated to an extra 10 to 20 days of sunny weather across the growing season; wheat and other cereals could be grown further north and at higher elevations. The warm climate, in other words, stretched the range of wheat even further north, farther from its original home in the Mediterranean foothills. I would like to tell the story of the amazing wheat plants that could grow anywhere and produce large, nutritious grains, but that would move my writing into the genre of speculative fiction. The information on exactly how wheat grew is scanty over time and space, but Marcus Terentius Varro, a Roman scholar from the last decades before the Common Era, noted that seeds grown in the fields of what is now Italy could yield 10 or 15 times the grain planted. For each seed planted, 15 would be harvested, so if a farmer kept one seed for planting the next fall, that left him with 14 to grind for flour.[5] (To put that into perspective, modern yields averaged across dryland and irrigated systems in North America are around 50 bushels harvested per bushel seeded.) It's clear that wheat, when moved outside its Mediterranean homeland, could grow across continental Europe during medieval times—but the yield in some years, instead of 15 seeds for each one planted, was closer to 3 or 4 seeds in parts of what now are England or France, while in what are now Scandinavia and Poland it was only 2 seeds grown for each one planted. The records from historical periods are scattered and rare, and not always straightforward to interpret. But we can conclude that while larger harvests supported a growing population, that increase resulted from more lands being plowed up and, to some extent, more grain being produced per field as farmers improved their practices and wheat adapted to new environments.[6]

Throughout this period, from about AD 800 to AD 1300, wheat was growing in North Africa, all around the Mediterranean Sea, and throughout Europe. The increased use of heavy plows meant that more lands could be turned into fields for farming, and cows and horses to help in that effort proliferated across the European continent. Villages grew, and with that the demand for food, along with labor to work the increasing number of fields plowed up for farming. Agricultural labor was always potentially limited for urban civilizations and growing populations. Shortages meant that laborers had leverage to negotiate their working conditions; during those times, serfdom decreased and landownership by peasants became more common. Apart from the brutal territorial takeovers and continued violence, things were slowly improving for wheat and for humans across much of Europe. But that changed. Unfortunately for the people living near the end of the time called the Medieval Warm Period, whatever climatic trend follows is likely to be a turn for the worse. In fact, what came next was the Little Ice Age, an aptly descriptive title.

In the early fourteenth century, the weather patterns throughout northern Europe were portentous for wheat and those people who depended on wheat. The summer of 1315 was so rainy that many fields weren't planted, because they never dried out enough to plant, or if planted, the crops rotted on the stem. A severe winter followed, and the summer of 1316 was even worse. The rains flooded fields and carried soils downhill and downstream. Across Norway, England, and northern France, wheat harvests declined between 25 and 50%. If in previous harvests each planted grain of wheat produced on average three or four grains and one of those needed to be saved for planting the following year, a decline of 50% meant that there was barely more grain harvested than was needed to be saved for the next year's seeds. And given the importance of grains to the peasants' diets, not to mention the physical labor that was an inherent part of the peasants' day, the shortage meant only one thing.

Hunger and famine weren't new to the people of medieval times, especially for those without land and power. Unlike the wide diversity of food sources that the early agricultural peoples accessed across the seasons in the ecologically rich areas along the Euphrates and the Nile, landlocked farming societies focused their diets on fewer foods. Grass and legume seeds had the advantage of withstanding storage, but even so, the stocks ran low or ran out before the next harvest. Families often relied on their vegetable garden and wild greens to tide them over until the new harvest's breads could be baked.

Besides those associated with the annual cycles of harvest and storage, food shortages can happen in any farming system, including modern industrialized farming, because of bad years, often the result of too much, too little, or poorly timed rain. Clergy and powerful families were better insulated from hunger than the rest of the population during food shortages, owing to social convention and better access to food sources. For the people of northern Europe in the early fourteenth century, repeated seasons without a harvest or a very poor harvest were enough to cause widespread starvation. Compounding the crop failures, the severe winters and an outbreak of cattle diseases throughout the rain-soaked lands killed half the dairy cows and oxen needed for plowing. Sheep stock declined by up to 70% in some areas, because of an increase in parasites and a lack of forage. The food shortages lasted until 1322; 10% of the population starved. Survivors were left with few options other than to move in search of food. People migrated south, abandoning flooded fields and farms, while the cities absorbed more people, even though the food supply there wasn't any better.

Famines produce an epidemic of starvation, and are usually the result of a combination of bad weather and political factors. The cold, wet weather of the early fourteenth century didn't impact the Mediterranean region like it did northern Europe, raising the question of why food wasn't transported to the north. In part, the grain production around the Mediterranean and in Egypt was feeding a populous Islamic civilization and thriving cities along the coast. The north could encourage an influx of food imports by paying higher prices for them. And in fact, the price of wheat in Paris rose by 800% as an impetus to draw in supplies from elsewhere. But the same rains that flooded the fields made the poorly developed roads equally inaccessible. The boats used to ship goods across the Mediterranean weren't suited to carrying grains up along the rough waters of the Atlantic coast, and the Pyrenees and the Alps were formidable barriers for horse-drawn wagons. The decentralized political system of feudal Europe wasn't organized to compensate for large-scale food shortages.

As serious as the famine was that struck down 10% of the population of northern Europe, what came after it was worse. The Black Plague killed as many as 50 million Europeans, or 60% of the population, between 1347 and 1352.[7] Some villages were completely wiped out, with no one left to bury the bodies. The winter weather slowed the spread of disease, but with the springtime warmth, more people fell ill, with some dying in just a few days. The worst of the plague in Europe occurred in the mid-fourteenth century, but the disease returned many times over the next few decades, each time killing fewer people. It would take 200 years for the population of Europe to recover to its size before the plague.

The drastic population decline in villages and cities meant a shortage of laborers. That was both good and bad for the plight of the poor. With that shortage, people could command a higher price for their labor. But when that price was deemed too high, there was an upshift in slavery. Landlords looking for people to work their farm fields enslaved Muslims and Jews from the Mediterranean region and North Africa. When additional labor was needed, orthodox Christians from eastern Europe were captured and transported to modern-day Spain to labor in the fields and factories. The demand for slaves declined only after European populations recovered.

The spread of wheat from the Syrian foothills throughout the Mediterranean region and then continental Europe paralleled the development of social stratification; those who controlled the land and therefore the production of wheat and other foods had the power. That power was used to accumulate wealth, built on the backs of the people who labored in the fields to turn the soils, plant the seeds, and harvest the grain. Then as now, the rich and powerful comprised a very small percentage of the population. And inequality and injustice were as evident to people of ancient times as they are today. Peasant uprisings and land reforms were less frequent, but as dependable and as ephemeral in their effects as the wheat harvests. Also present in European farming was the stratification of food production for the wealthy versus the working classes. If peasants couldn't afford to use their own fields to grow grains that may not be successful, they were required to tend those same fussy wheat fields for the nobility, to provide the grains that could be ground into white flour. This flour was used to make a bread that could soak up the juices from the nobles' meat dinners—typical fare for them, but not for the laborers.

Farmers growing food to sustain their families chose their crops based on the climate. The Little Ice Age lasted from the fourteenth to the nineteenth century. During the weather fluctuations between cooler and warmer periods throughout Europe, many of the peasants and small farmers favored rye and oats over wheat, because they grew more dependably in the cold weather. Years with cold winters and rainy summers continued to destroy harvests, and famines and hunger were part of the fabric of life. But life in Europe did change, albeit slowly: from feudal classes to the beginning of nation-states; from a largely agricultural economy to the beginnings of the Industrial Revolution. The intensification of food production happened hand in hand with the population shift from the rural landscape toward urban areas. That shift occurred in part because the Industrial Revolution required a large labor force. Europe morphed from a largely agrarian lifestyle, with 90% of the people working in farm fields at the beginning of the medieval period, to a heavily industrial lifestyle, with 50% of the people living and working in villages and cities at the beginning of the nineteenth century. Intensified farming was needed to support the larger urban population. This period also saw an increase in farming for profit, usually with a narrower scope of production than the subsistence farming that provided all the food and fiber for the farming family.

We all have times in our lives that are less inspiring than others, so perhaps wheat can be forgiven for a relative lull in its existence. During the centuries of cold and wet weather, most of the European farmers turned their back on wheat, given that its yield was unreliable in the fickle climate. But the nobility still required their light breads, and indeed it was the nobility who stepped in to bring on the next adventure for our struggling hero. The world was opening for Europeans, who used their finances and growing expertise in ocean navigation to claim new lands for the royalty. This age of exploration was a time of expansion for Europeans, and the opposite for the peoples dwelling in the lands about to be discovered. Whether it was the result of the explorers' weapons, or the diseases they and their animals carried, or their treatment of the people they enslaved, European imperialism caused the decimation of the indigenous people in the new lands. Wheat was part of the imperialistic story, as its seeds often traveled with the explorers to ensure that they would have familiar foods in their new home.

The quest for new lands and riches inspired Europeans, initially the Spanish and the Portuguese, to travel afar, crossing the Atlantic in search of treasures. One of those voyages was made by Christopher Columbus, who embarked in 1492 with the *Niña*, the *Pinta*, and the *Santa Maria* in search of a trade route to Asia. Instead, he landed in the Caribbean, left 35 of his men there, and returned to Europe to gather more resources. He came back with 17 ships filled with 1,200 men, sugarcane plants and wheat, the seeds of fruits and vegetables, and an assortment of horses, pigs, and cows. His introduction of animals and crops from Europe was a big disruption to agricultural systems that had been developed for soils and climate of the Americas.[8]

But European colonizers had little interest in New World approaches to farming. Instead, they imported their own farming tools and livestock. They used the expanded agricultural reach across the Atlantic to produce sugarcane, coffee, cocoa, tobacco, and indigo. Instead of the careful rotations and management of farmland to maintain soil fertility over the long term, practices back home that kept people fed with varying degrees of dependability, they developed an agricultural system in the New World based on growing single crops over large areas. Then they exported those foods to the developed countries and eventually to new colonies in North America. In the seemingly endless expanse, lands could be farmed until the soils were exhausted. At that point, all that needed to be done was to clear new land—as long as the colonizers had no regard for the indigenous people living there.

The people native to the New World were either enslaved to work on the colonists' plantations or pushed to areas with less favorable conditions. The Europeans arriving in Central and South America came for new opportunities for wealth. Rather than organizing farming systems to feed local people, the colonists made their farm fields the source of luxury goods for the increasingly prosperous in Europe, who could afford to nurture a desire for specialty items. Those merchants who oversaw the growing and shipping of these goods to Europe became wealthy on the backs of slave laborers in the fields. Moreover, slaves were traded along the western coast of Africa and eventually the eastern coast for cotton, gold, silver, and rum. The commerce in human lives was as lucrative as that in the sugar and tobacco made possible by those same people.

To attract European settlers, the New World needed something of a makeover so that it would feel more familiar to these new arrivals. Spaniards, for instance, brought wheat, oats, barley, rye, broad beans, alfalfa, grapes, olives, and fruit trees to Peru. They also brought oxen, pigs, sheep, horses, donkeys, mules, and poultry. Various tools made the journey from Spain as well: sickles, hoes, spades, ards, grain mills, and carts. Wheat crossed the Atlantic, but the tropical climates of the New World were no better for this Mediterranean grass than were the monsoon climates of India. It could grow only in the higher elevations and across some of the less humid pampas. By the mid-sixteenth century, wheat was harvested in the highlands of Mexico and Central America, the temperate areas of Peru, and the highlands of Chile.

In the Northern Hemisphere, wheat arrived with the Puritans in the 1660s. Its seeds, adapted to British winters, couldn't survive the harsh New England winters, so settlers switched to planting wheat in the spring. But a spring planting delays the harvest till later in the summer, leaving the wheat crop vulnerable to summer storms, insect attacks, and fungal pathogens. Sure enough, in 1665 most of the wheat was ravaged by black rust, one of the fungi that plagues wheat fields. The first colonists grew very little wheat, but it wasn't for lack of trying.

Although some of the settlers saw the New World as a place to build a community and raise families, others saw the potential there to tap into the rapidly expanding world economy and provide food and luxury items for the established and growing cities and nations of Europe. Planting blocks of single crops across large regions was possible in the New World, an area without Europe's rules and conventions. Agriculture developed across North America, with regional differences based on climate and culture. The dense forests along the New England coast represented a wild force to be tamed, so the farming established there was primarily subsistence farming. The Southern states adopted the logic of plantation farming, with intensive labor requirements met by importing African slaves; tobacco and cotton were produced with slave labor. But the prairies of the Midwest and the rugged plains of the West, with fertile soils and plentiful sunshine, were the perfect opportunity to grow wheat and corn as far as the eye could see. And the cost of plowing up the new land and setting up farms was funded in part by those with money to buy into these endeavors. The westward expansion across North America relied on investors, and changed the nature of agriculture from the subsistence farming of the New England settlers to farming for the market as a way to repay loans needed for the land, the machinery, and the labor.

Wheat's unimpressive performance in the New England climate came as no great surprise. What's impressive about its move across the Atlantic is that once it arrived, it eventually spread across the entire North American continent, grown at one time or another from the East Coast to the West Coast, from Canada to Mexico. But that capacity of wheat to grow across broad latitudinal and longitudinal gradients is actually the result of its many different transatlantic migrations.

The grains of wheat that the Puritans carried to North America were bread wheat, the species formed by hybridization 8,000 years earlier in a field of emmer wheat. But by this time, bread wheat had differentiated into different varieties—still all the same species, but each with unique characteristics. There was spring wheat and winter wheat, red wheat and white wheat, hard wheat and soft wheat. Winter wheat is a winter annual, germinating in the fall and growing until the weather gets too cold, overwintering as a small plant, and ready to grow again in the spring as soon as the temperatures warm up. Winter wheat not only tolerates cool winters but needs the cold to trigger its reproduction. In contrast, spring wheat doesn't need a cold period to reproduce. It can be planted in the spring, but compared with fall-planted varieties, it requires a few extra weeks in the summer to mature. When people planted wheat in areas with bitterly cold winters, the fields didn't always green up in the spring. If a farmer had seeds left over from the fall planting, he could try spring planting. Those plants that survived and grew to maturity without the need for a cold period were spring wheat. Spring wheat has the advantage of not having to endure the cold winter months in order to mature. But winter wheat has a couple of advantages, including an overall higher production level, because the wheat grows longer between the fall and spring growing periods. And an early harvest means a greater chance of avoiding the summer hailstorms that can damage entire fields.

White wheat and red wheat differ in the color of their bran. White wheat is like an albino version of red, its bran having lost color, along with some tannins and flavonoids, giving the white wheat a somewhat softer flavor. Hard wheat and soft wheat differ in the amount of puroinoline, a protein produced by a gene donated from the second goatgrass that with emmer had created the hybrid bread wheat. That gene confers softness to the wheat kernel, resulting in a grain that is easier to grind, with rounded edges as opposed to sharp corners on the flour. Hard wheats are the product of a mutation in the hardness gene, so less puroinoline is produced; hence hard wheat more closely resembles emmer and durum wheat. Hard wheats are used for breads and rolls, while the soft wheats make light cakes and pastries and crackers.

The early colonists in Massachusetts grew a soft wheat, likely a soft winter wheat. By the mid-nineteenth century, wheat in North America was being grown primarily in the Midwest, and it was all spring wheat. Winter wheats couldn't survive the cold winters, but the spring wheats weren't very productive. This impasse was overcome with help from some Canadian ingenuity. David and Jane Fife in Ontario had received some wheat seeds from the city of Gdansk, on the Baltic Sea. Only one of the handful of seeds they planted in the spring matured into a grain-bearing plant, albeit with just five seeds. The story goes that their cow ate two of the seeds, but they saved the other three to plant the following spring. With patience, skill, and luck, they increased their seed yield each year until they had enough to plant their fields and share with other farmers. That wheat, eventually grown across much of Canada and introduced to Wisconsin, Minnesota, and the Dakotas, became known as Red Fife. It was the first hard spring wheat in North America, and its use spread across North America because it produced well.[9]

Another important voyage for wheat came a few decades later, when it was brought to the United States by a group of Russian Mennonites. Because of their reputation as good farmers, the railroads had recruited them to colonize the southern midwestern plains. In preparation for their emigration, they collected seeds from the hardiest wheats. One variety, named Turkey Red because it had been brought from Turkey to southern Russia in 1860, traveled with the Mennonites to the United States in 1873. By 1919, Turkey Red made up 30% of the wheat grown in the nation, and almost all the hard winter wheat grown.[10]

Besides crossing the Atlantic with the European settlers, wheat moved to Australia and New Zealand in the early and mid-nineteenth centuries, respectively, with immigrants who struggled to get their first crops established. The Chinese have been growing wheat for at least 5,000 years, and they are the likely source for its spread to Japan and Korea over 2,000 years ago. The Silk Road, an ancient trade route network, was one of the paths for wheat to enter India, where it has been grown for several thousand years. Wheat has been in Ethiopia for thousands of years, probably moving inland from somewhere along the Nile. The Dutch brought wheat to South Africa in 1652. And in the nineteenth century, British missionaries brought wheat to Kenya, while German missionaries brought it to Tanzania near that century's close.

As wheat traveled around the globe, the style of farming changed when Europeans crossed the Atlantic for the New World. In fact, wheat yields were higher in Europe than in the Americas. European farmers who remained in their homeland had access to more labor. Their fields were confined by the growth of villages and cities and industrial areas, so there was more of a focus on maintaining soil quality and nutrients for better crops. But when American grain started to be shipped across the Atlantic at prices lower than what Europe's labor-intensive farming methods could compete with, Europe's wheat fields contracted.

Like today, many of the people who depended on their fields to feed themselves and their families were hardworking and innovative. But that's not to say that our ancestors' relationship with wheat was pure bliss. Farming has always been hard work, but the application of iron and later steel implements increased the kinds of lands that could practically be cleared for farming, expanding the acreage for cultivation. Improved tools and the introduction of livestock were important breakthroughs for increasing food production. However, it was the advent of industrialization in the eighteenth century that lessened the demands for human labor. But even with better tools, farmers struggled with changing climates and severe weather patterns. Relationships are hard work, but our species never gave up on wheat.

CHAPTER SIX

Nurture and Nature

From foraging for grass seeds, to cultivating fields, to trading large quantities of grain for both necessary and luxury items, the ability to grow wheat and grow it well was for some a matter of life and death, and for others a means to amassing a fortune. As populations increased, so did the scale of agriculture. But regardless of whether farming took the form of a small subsistence effort or fields managed by the political state, it was never far removed from the realities of environmental variation and its effect on wheat and other crops. During cold and wet periods or hot and dry summers, crops failed. And with the crop failures came the stark realities of famine and starvation.

The challenges and the necessity of growing food are shouldered by farmers but are also the burden of the state. Political

unrest derives from many circumstances, but shortages of food, or the corollary—soaring food prices—inspire people to take their dissatisfaction to the streets. One way for political powers to ensure steady food production is to control the markets so that consumers can always be assured of affordable food. Another way is to invest in programs that promise to improve farming methods and disseminate that information to farmers. And that brings us to the origins of agricultural science, focused on the application of scientific methods and experimentation to improve agricultural outcomes.

Early advances in farming technology (ards and moldboard plows) and farming practices (crop rotations, compost and manure additions) helped increase the land area where crops could be planted, as well as enhance the productivity of those fields. The early farmers used their observations and natural or planned trials to understand what worked and what didn't, and that understanding got passed down from one generation to the next: sometimes etched into mud tablets, later stored in the form of accounts or written summaries, and much later published in farming newsletters and magazines. The philosophers and scholars, for most of history a rarified group, interpreted the world through the intellectual lens of their time. They recorded their own view of the world, which was made available to all who could read their texts—which for most of history excluded the farmers. The ties between the science and the practice of agriculture have always been tenuous.

Consider, for example, what it takes to grow a field of wheat. Sunlight and carbon dioxide come together in a plant's leaves to build the carbon backbone that structures the plant. Its roots take up water from the soil, and carry the nutrients dissolved in the soil water into the plant. Plants need these nutrients to build the machinery that drives photosynthesis, and to form all the structures for flowering and producing seeds. Without understanding these details of plant physiology, our ancestors were successful farmers. The Romans fertilized their fields with manure, ashes, cinders, and cut lupine plants. The understanding at the time, based on Greek scholarship, was that plants grew because they could absorb a vital force from the organic matter in the soil that allowed them to grow. The manure and cut lupine plants included fragments of previously living plants and animals that contained the vital force, and its transfer caused the growth of the field of wheat. Up until the nineteenth century, a vitalist theory of plant growth dominated European thinking related to agriculture.

While farmers struggled to optimize field rotations and respond to weather fluctuations, early scientists were puzzling over how plants grow. Jan Baptista van Helmont, a Belgian man of letters born in the late sixteenth century, was one of the first to test the idea that plant growth depended on its extracting the necessary forces from soil organic matter. He weighed a young willow tree (it was 5 pounds) and then planted it in a pot holding 200 pounds of soil. For 5 years, van Helmont tended the willow, providing either rainwater or tap water but nothing else. At the end of the experiment, the willow weighed 167 pounds, and the soil was just 2 ounces below its original weight. From these observations, van Helmont concluded that plant growth cannot be based primarily on what a plant took up from the soil. He surmised that something in the water somehow transformed into plant material.

It took another century of refining methods for measuring compounds, particularly the compounds in the air, before scientists could better interpret van Helmont's work. A series of experiments over the years proved that the atmosphere is a mixture of compounds, and that mixture includes oxygen, carbon dioxide, and nitrogen. A Dutch scientist, Jan Ingenhousz, is credited with discovering photosynthesis, by showing that plants take up carbon dioxide and release oxygen, but only in the presence of sunlight. Even with the first rudimentary idea of photosynthesis, there was still the prevailing idea that the humus in the soil—that dark, partially decayed organic matter that gives a rich topsoil its hue—provided the vital force necessary for plant growth. That idea was supported by practice, in fact, as farmers for millennia had seen that adding manure and compost to soils increased crop growth.

Two German chemists active in the nineteenth century, Carl Sprengel and Justus von Liebig, were instrumental in advancing scientific knowledge beyond the humus idea of plant growth to parse out the requirements for that growth from a basis in chemistry. Liebig is often credited with being the first agricultural chemist; but Sprengel, 16 years his senior, had taught the first university course in agricultural chemistry years earlier. Liebig is credited with coming up with the law of limiting factors, which describes the relationship between plant growth and all the elements used for building plant tissue; but Sprengel had published the same idea 27 years before Liebig. That law is based on the understanding that the elements in plant tissues are necessary for plant growth. Sprengel had grown plants in different soil types—some rich with humus, hence with the potential to transmit the vital force to the plant, and others in mineral soil, with no organic matter. Plants grown in either soil type had the same elemental composition, suggesting a couple of things. First, the theory of transfer of vital forces was pretty much discounted; plants grew well even if no organic matter was in the soil. And second, the elements necessary for plant growth could be taken up from mineral soil. Therefore, if one knew which elements a plant needs to grow—and with advances in chemistry, chemists now could measure those elements—and which elements were present in the soil—also possible with advances in chemistry—one could add what was needed to the soil to optimize plant growth. Farmers had been doing that by adding manure and compost, but in an increasingly crowded world where villages were becoming cities, and industry was taking over the places where animals once grazed, a shortage of manure and compost could be compensated for with mineral fertilizers.

Liebig, who had a great propensity for self-promotion and publication, summarized his thoughts in a book first published in 1840, *Organic Chemistry in Its Applications to Agriculture and Physiology*. A revised edition came out every year for a few years as Liebig's ideas changed. In the earliest editions, he emphasized the importance of nitrogen for plant growth, and minimized the need for the rest of the elements that comprise plant tissue. But by the third edition, he reversed his recommendations, laying out the significance of adding minerals to soils to increase plant growth. Nitrogen, he thought, being a major component of the atmosphere, was likely taken up by plants without any need to add it to the soils.

But plants don't take up nitrogen from the atmosphere. Only a few species of bacteria can do so; and only a few species of plants, mostly the legumes, house those bacteria in little nodules on their roots. One of the first mineral fertilizers was saltpeter, nitrogen-containing salts that form naturally in soils from the microbial breakdown of organic matter.[1] In dry climates the salts, instead of dissolving in water, precipitate, meaning they come out of solution, and form white crystalline layers on rocks or stable walls. Once precipitated, the salts are easily scraped off the surface and collected. Saltpeter was also extracted from compost heaps containing dung, and was so easily derived that an instruction manual about that published in 1862 reports that "every Swede pays a portion of his tax in nitre"—another term for the nitrogen salts.[2] While saltpeter was recognized as an effective additive to farm soils, it was rarely used as a fertilizer. It was the primary ingredient in gunpowder, and for some reason ammunition manufacture was a higher priority than food production. Other inorganic fertilizers used were chalk and marl, both forms of limestone, in addition to common salt. Limestone is calcium carbonate, and thus can be a source of calcium and reduce soil acidity.

Interest in fertilizers had been increasing in Europe when, in 1804, Alexander von Humboldt, a Prussian-born naturalist who traveled extensively on natural history expeditions, returned from Peru with a sample of guano. The Incas used guano, the deposits of bird dung on coastal islands, as a fertilizer. Consequently, they protected not only the islands where the guano was deposited but also the seabirds themselves. In dry climates like coastal Peru, the bird droppings, rich in nitrogen and phosphorus, accumulate on rocky islands. In wetter coastal areas, such as around Great Britain, rainfall dissolves first the nitrogen component and then the rest of the droppings, so whatever guano remains has much lower fertility. British and later French entrepreneurs paid the Peruvian government to mine the guano, brought in slaves from China to process it, and shipped it across the Atlantic or, to a lesser degree, to North America. The guano deposits were depleted within 50 years, but the demand for fertilizers only grew.

The next nitrogen source to be tapped was the sodium nitrate deposits of the Atacama Desert in South America. Wedged between the Andes Mountains and the Pacific Ocean, this desert is one of the driest areas on earth outside the polar regions. Some parts receive less than one-hundredth of an inch of rain per year; the wet spots receive a little over half an inch. After Peru and Bolivia warred against Chile, the desert was ceded to Chile, which recognized the value of the sodium nitrate deposits. In fact, those deposits were in such high demand that by 1920, 46,000 workers were toiling in the deserts of Chile. The mining continued until scientists figured out how to mimic the microbial process in factories. Two German scientists, Fritz Haber and Carl Bosch, came up with a way to fix nitrogen industrially (Haber) and then to employ the method on a scale that would enable the factory manufacture of fertilizer (Bosch). Instead of burning the storage molecule ATP to break the triple bond of gaseous nitrogen with the help of an enzyme, as the single-celled bacteria are so adept at doing, pressure—200 times normal atmospheric pressure—and high temperatures—800 °F—were used to combine the nitrogen and hydrogen to form ammonia. Once that process could be managed, the sodium nitrate mines of Chile went quiet, although mining of gold and lithium continues there today.

Besides access to nutrients as a key influence on its productivity, a field of wheat grows well or poorly depending on weather. The success of wheat in North America was partly the result of discovering varieties that could grow well in the new climates. Transporting crops to new areas was just one way that we have modified the evolutionary trajectory of our food species. Agricultural manuals from Greek and Roman times advised setting aside seeds from the largest ears for next year's planting (selecting for higher yield potential), and gathering seeds from distant sources to look for variants that might grow bigger or taste better (adding variation to the population). These days, we argue over whether transferring genes from one species to another is safe, when in fact the greatest changes inflicted on our food species were made by our ancestors during the early domestication of our food plants. At that time, they had no understanding of the mechanistic basis of how plants grow and pass on traits to the next generation.[3]

Plant breeding entails the deliberate identification and selection of traits by humans, so it's different from the inadvertent selections our ancestors had been making for thousands of years. Although breeding is based on an understanding of the evolutionary processes of inheritance and selection, early understanding of those processes was in fact aided by the practice of breeding. Selectively breeding animals is one thing, but the application of those same principles to plants requires an understanding of the anatomy of flowers—knowing that their pollen carries the chromosomes from one plant to combine with the chromosomes in the ovule of another plant in the act of fertilization. The first recorded intentional pollination comes from Mesopotamia and Egypt and involved date palms. Many plant species produce both the pollen and the ovules in a single plant; but date palms, like gingko trees and asparagus, have some plants that produce only pollen and others that produce only ovules, the equivalent of male and female plants. Ancient stands of date palms were planted with just a few males scattered among mostly female palms to maximize the number of dates produced. That pollen functions as the male principle in flowers was rediscovered and articulated in Europe by the end of the seventeenth century, but widely held societal beliefs there slowed the development of breeding programs. Intentionally breeding plants to select for specific traits was seen as an intrusion on God's will; that which God created was not man's to destroy.

Then there's pigeons. At first glance, it's not easy to imagine what einkorn, emmer, and bread wheat have in common with fantail pigeons. Both the peacock-like tail that merits the name for these birds and the walnut-shaped protrusion on the upper beak of passenger pigeons were traits highly prized by nineteenth-century gentlemen who bred pigeons for show. When these gentlemen found a trait they liked on a bird, they bred that bird with a close relative in the hope of producing more pigeons with that same trait. This practice, known as inbreeding, had been taboo in human societies for thousands of years in order to prevent defective offspring, the result of their higher likelihood of inheriting rare, recessive traits carried by their parents. But our standards for humans and pigeons differ; the inbred pigeons that carried the desired tail feathers were saved, and those that had some other malfeasant trait could be sacrificed, all in the name of a spectacular new pigeon breed.

Variation in pigeons was an inspiration for part of the arguments put forward by Charles Darwin to explain the process by which species change through time. The mechanism behind the change was natural selection, which Darwin derived from two premises: the prolific variation found in nature and the struggle to survive. Because his ideas were revolutionary in the late nineteenth century, when most people accepted the role of the Divine Hand in the creation of life, Darwin carefully laid out his arguments to counter what he knew would be strong public opinion against his ideas.

Darwin's first premise of natural selection is that within any species, individuals differ from one another. Those differences may be minor, but the variation is an intrinsic part of nature. This premise is difficult to construe as controversial. However, at the time the accepted idea about species was that of immutable forms—species created by God did not change through time. So Darwin constructed his argument beginning with the least controversial example: pigeons. Domesticated over 5,000 years earlier, these birds were adopted as pets, used as religious icons, and trained to deliver messages and aid navigation. The traits considered valuable among British gentlemen were diverse. The fantail pigeon has so many tail feathers—30 to 40 instead of the dozen found in most pigeons—that its tail stands up like a peacock's. The pouter pigeon has a flexible and enlarged esophagus and crop that it can inflate, resulting in a balloon-like bulge below its beak. Tumbler and roller pigeons are named for their self-explanatory flight style, and trumpeter and laugher pigeons for their unique calls. The pigeon varieties bred by the British gentry differed in the number of vertebrae, the curve and length of the facial bones, the number of wing and tail feathers, the size and shape of eggs, and the manner of flight. Yet all the varieties derived from the same original species: the rock pigeon, *Columbia livia*. Given that so much variation can arise from a single species, then why, Darwin argued, is it not possible that species in nature are lineal descendants of other species? "Descent with modification" was the term that Darwin used for what we refer to as evolution.[4]

The next part of the argument extends the documentation of natural variation to nondomesticated species. The naturalists of the time used a fair amount of ink describing and arguing over the taxonomic status of plants and animals observed on their forays into nature. What one eminent scholar called a species another called a variety or a geographic race. Darwin considered the distinction between the taxonomic categories "entirely vague and arbitrary." The British red grouse, for example, was considered by some eminent ornithologists a race of a Norwegian species of grouse, while a long list of other eminent ornithologists classified it as a separate species. In light of such a fuzzy distinction, Darwin suggested that the well-marked variety could just as easily be called an incipient species. In contrast to species as immutable forms, his view of nature was that of a dynamic, changing system.

The second premise of Darwin's argument, the struggle for survival, was influenced by Thomas Malthus's *Essay on the Principle of Population*, written at the end of the nineteenth century. Malthus, a British economist, described the stark mathematical realities of the difference between an exponential growth rate for humans and an arithmetic growth rate for our food supply. Basically, while human populations have the capacity to double, then double again, and continue to double, the food supply increases at a steady increment. Given these conditions, Malthus argued that the number of humans will increase to a point beyond sustenance, and only famine and vice will keep the population in check. The long list of famines throughout Europe caused by variable climate and poorly connected trading networks was the context for Malthus's thoughts.

Darwin transformed Malthus's ideas of starvation and disease as natural population controls into what he termed the struggle for existence. That struggle is a natural consequence of the high rate of reproduction of all living beings. For example, consider a plant that lives just one year and produces two seeds before it dies. Suppose both of those seeds germinate and grow into a mature plant and produce two seeds before they die. After 20 years, there will be one million individuals of that annual plant. Likewise, if you start with two elephants and apply a conservative estimate of first breeding at the age of 30 years and then two more times by the age of 90 years, within just five centuries there would be 30 million elephants. But the world isn't overcrowded with elephants, and except for our agricultural fields, plant communities are usually composed of many species. Clearly, plants and animals realize only a small fraction of their reproductive potential. Many of those seedlings and elephants don't live long enough to produce more seeds or more baby elephants. Therein lies the potential for natural selection.

If that small fraction of plants or animals that survive to reproduce isn't a random selection but is instead based on which individuals are more closely adapted to the trials and variations of their environment, then this is "descent with modification," based on heritable variation and environmental forces. Those environmental forces act as the grand shopper, deciding who goes into the grocery cart that leads to future reproduction, and who stays on the shelf to languish. Starting with breeders' capacity to select traits in pigeons, and understanding that nature can exert a similar kind of selective force, Darwin laid out the basis of his theory of natural selection. The difference between natural selection and selective breeding is simply the force selecting for particular traits. The pigeon fanciers had a specific end goal in mind. Environmental variation, be that cold winters, wet summers, or insects that lay eggs in the seed head, is a bit like having 10 people cutting your hair at one time and hoping it will look good in the end. Environmental pressures are varied and come from climatic and geologic factors along with interactions with other organisms, be they competing plant species, fungal pathogens, or hungry animals.

Heritable variation in a population that results in differential survival and reproduction is the basis for natural selection. While Darwin laid out his arguments based on years of careful observations of plants and animals and insects on his voyages around the world and in his backyard, there was one nagging hole in his argument: how are traits passed from parent to offspring? On a large scale, it was easy to see that children have a blend of traits from both parents. Likewise, pigeon breeders selected birds with desirable traits to produce offspring with those same traits, and a long line of people who raised livestock knew to mate the biggest, strongest bull with a cow that had equally promising traits. But the result of those crossings wasn't always predictable. To control the outcome of your crossings, you must understand the mechanisms underlying inheritance of traits. And that's where the work of an Austrian monk comes in. Gregor Mendel was conducting large and carefully designed experiments while Darwin was summarizing the observations from his travels. Mendel spent years on a series of meticulous experiments in the gardens of his monastery, although it was decades after those results had been published before his work was finally read and understood.

Mendel was prone to scholarship, and was specifically interested in a topic deemed controversial in light of the religious doctrine which held that species were immutable. Mendel worked with pea plants, following in the footsteps of a long line of experiments conducted by English gardeners. These experiments showed that if someone crossed two pea plants that had different traits, the next generation would have a mix of traits from both parents. It took until the third or fourth generation before plants bred true, meaning that the next generation had the same traits as their parents. Mendel built on previous studies with a series of carefully designed experiments in his monastery garden that gave insight into how genetic variation was passed from one generation to another.

The pea plants that Mendel worked with had specific traits that had been studied by other scientists—tall or short, white or purple flowers, smooth or wrinkled peas, and yellow or green peas. By fertilizing one plant with pollen from another, keeping careful records of the traits of each pea that resulted from those fertilizations, and then planting the new peas and fertilizing each plant with known pollen, he accumulated a set of data which helped him describe how traits are inherited. His results were interpretable because those traits, we later came to understand, were coded for by a single gene. Different forms of a gene are called alleles, so whether a plant produced green or yellow peas was the result of what allele that pea had inherited from each parent. From the inheritance patterns he recorded, Mendel identified some characteristics as dominant traits and others as recessive. A dominant trait is expressed when the plant has inherited either a single copy of the allele coding for the dominant trait from only one parent or a copy of that allele from both parents. Likewise, recessive traits are expressed only when both alleles that a plant has inherited code for the recessive trait. For his experiments, Mendel used seeds of plants which bred pure, meaning that a parent with yellow peas always produced progeny with yellow peas, and likewise for plants with green peas. When a purebred line of yellow peas (described as YY, meaning that both alleles inherited from parents code for yellow peas, a dominant trait as indicated by the capital letters) is crossed with a purebred line of green peas (yy), all the seeds produced by that cross have inherited a yellow gene (Y) from one parent and a green gene (y) from the other parent, so the shorthand description for the new pea is Yy. Those seedlings will develop into pea plants with yellow peas, because yellow peas can be either YY or Yy. Green coloring, in contrast, is a recessive trait, requiring a copy of the green pea gene from each parent. All green peas are yy.

Mendel kept careful records of the plant traits in each generation, and he made thousands of crosses, tending around 29,000 plants over the years. With this work, he was able to deduce principles that are the foundation of modern genetics. First, he surmised that traits are passed on from parent to offspring as a result of inheritance factors. These are what we now know as genes. Second, individuals inherit those factors from both parents, not just from the pollen, as Aristotle believed, or just from the ovule. And third, those factors may not be expressed in the parent, but they can still be passed on. For example, when Mendel crossed the first generation (the Yy generation, with just yellow peas) with itself, most of the plants had yellow peas (either YY or Yy), but a quarter of them produced green peas (yy). This was part of the explanation for why crosses didn't always turn out as expected, based on the traits of each parent.

Mendel published his results in a scientific journal in 1866 and sent copies of his article to leading scientists, including Darwin, whose copy remained sealed and never read. In fact, it was 16 years after Mendel's death when scientists discovered his work by reading his previously published monograph and realizing he had anticipated their discoveries by 30 years. Mendel made a large contribution to something that Darwin puzzled over in his work—a statistical proof of transmission of genetic factors from one generation to the next. As noted earlier, what made Mendel's work possible to interpret was that he chose traits that were controlled by a single gene. Many if not most traits are controlled by multiple genes. Height in humans, for example, is controlled by about a thousand genes. We used to think that eye color was coded for by a single gene, the dominant version being brown and the recessive blue. But now we understand there are 16 genes that code for eye color, which is why two blue-eyed parents can sometimes have a brown-eyed child, or one person can express a different color in each eye.[5]

Once Mendel's research was discovered, it took some time before it was integrated into traditional breeding programs. That's because most of the traits that plant breeders are interested in are controlled by many different genes. Like human height, how much seed a crop can produce is affected by many different factors, including the rate of growth, the timing of the initiation of flowering, the growth of stem versus leaves, and the amount of carbohydrates packed into the seed. Even for traits controlled by many genes, each individual gene is inherited just like those Mendel studied. But unlike a trait controlled by a single gene, a trait affected by many genes responds incrementally, the result of the combination of the two alleles inherited from each parent for tens or hundreds of different genes. And then just to keep it interesting, for bread wheat, with six instead of two copies of all the genes, there's a little more sorting to do. For plants that are polyploids, meaning they have two or three or more copies of the whole genome (a living being's genetic material), some genes are expressed in each set of chromosomes, and some genes are turned off or even deleted in one set of chromosomes. It makes breeding while thinking about the genetic mechanisms a bit more complex.

The units of inheritance that Mendel described were genes, but at the time that Darwin and Mendel were writing, scientists weren't yet using that term. Another 100 years would pass before we understood that it was DNA that was the hereditary material. Chromosomes, in fact, were first visualized in the lab under a state-of-the-art microscope in 1882. That chromosomes carried what Mendel had termed the inheritance factors was supported in the early 1900s, with the observation that body cells carry chromosomes in pairs, while in the reproductive cells (eggs and sperm in animals and ovules and pollen in plants) only half the chromosomes are present—a single copy from each pair. While one of our cheek cells holds 46 chromosomes, there are only 23 chromosomes in our egg or sperm cells.

Genes are the unit of inheritance, and a simple way to envision them is like pearls on a necklace: the necklace is a chromosome, and each pearl is a gene. Chromosomes are two strands of DNA that elegantly spiral around each other in the form of a double helix. That double helix is composed of four base pairs—adenine, cytosine, guanine, and thymine. They are base pairs because each time adenine is on one DNA strand, thymine is directly across from it; correspondingly, cytosine is paired with guanine. The order of the bases on a single strand of the double helix is what determines the function of the gene, as different combinations of base pairs, in sets of three, encode for a specific amino acid. Amino acids combine to make proteins, and proteins regulate much of the function of the cell. Understanding the genetic code—from the structure of the DNA molecule, to the composition of the base pairs, to the process of transcribing DNA into RNA, and then translating RNA into amino acids—is one of the more breathtaking aspects of genetic inheritance.

This sketch of how inheritance works is pretty straightforward so far, thanks to omitting many of the complicating details. One such complicating detail is that only about 2% of the base pairs on a chromosome are genes. To make the analogy of the pearl necklace slightly more accurate, then, we would substitute a necklace made by a young child: plastic beads and all sorts of unknown things strung on a line, and every now and then a pearl. And even that analogy isn't exactly right. The genes, or pearls if you will, are usually broken up by pieces of noncoding DNA, so it's as if the child had smashed the pearls with a hammer, inserted pieces of Lego bricks or tiny pebbles into the pearl rubble, glued it all together, and then strung it on the necklace. Since the 1970s, we called the 90-something percent of the DNA that wasn't genes "junk DNA." That term has mostly fallen out of use these days, as we slowly uncover the function of the DNA that doesn't code for genes. With each new study, scientists incrementally build on our understanding of how our DNA functions within the context of the cell and the rest of its machinery.

Our interest in breeding had begun with animals, long before we understood mechanisms of inheritance. Perhaps incorrectly, Roman instructions for breeding oxen mostly ignore the contribution of the bull to the progeny's traits, but provide a list of attributes to watch for in the females, including a strong torso and buttocks, well-shaped feet, clear eyes, and a healthy coat. Horses were similarly bred for racing, and donkeys and mules as important work animals.

As for plants, while our ancestors influenced their evolution by selecting, for example, the biggest berries, the plants that were the easiest to harvest, and the fruit from the trees that tasted best, it was much later that we did intentional crosses. In the early nineteenth century, a new practice of selecting a single plant, and maybe a single seed head of that plant, was the start of a practice that led to the development of elite lines. Generated from a single plant, those lines produced seed that had more predictable performance for the farmer, and included a lot less overall genetic variation in a single field.

The demands of growing wheat across wide temperature and precipitation gradients throughout North America were the foundational test for the new science of breeding. Initially, our breeding programs were based on trial and error, testing wheats from different regions to learn what grew best in a new region. In 1840, 41 varieties of wheat were being grown in New York State alone; similarly, Ohio registers include 111 varieties of wheat in 1857. Some varieties grew better on certain soil types, others in specific climates, and some were more resistant to pests. With all those varieties, it's clear that the ones we still talk about are the Marilyn Monroes of the wheat world. Red Fife from Canada has been one of those stars. As a result of its capacity to be planted in the spring, important in regions with winters too cold for survival of fall-planted wheats, farmers across the upper Midwest planted Red Fife. Turkey Red wheat, a winter wheat brought to the United States by the Mennonites, expanded our capacity to grow wheat in the cold (but not brutal) winters and dry summers of Kansas. And then there's Marquis, the product of a cross between Red Fife and red Calcutta wheat. Marquis accounted for 90% of the Canadian wheat harvested in 1920, and at the time was grown from Washington to Illinois.

To optimize the growth of wheat and other crops, breeders select the individuals from a cross that inherited the most desired traits from each parent. Wheat naturally self-fertilizes, that is, produces offspring when the pollen and ovule come from the same plant. Thus, once identified, a superior plant can produce seed for a couple of generations, and all the seeds will carry the same genes as the original selected plant. The difference between a field of wheat in the eighteenth century and a field that was the result of breeding and selection is striking. The yield is much lower in the eighteenth-century field, but the genetic variation is much higher. The eighteenth-century field is representative of the landraces growing in scattered fields in a few corners of the world where farmers aren't growing commercially produced cultivars. Landraces are valuable because the variation that breeders got rid of, the smaller, scraggly, less productive individuals, may also hold the genes for greater tolerance to abnormal rainfall or late frost or fungal pathogens.

And this is the great oxymoron of plant breeding. Breeding is the intentional selection of superior variants in a population. But once we select for plants with the best trait, we eliminate other genetic variants, leaving less variation from which to select in the future. A constant challenge for breeders, whose job is to select the variants that are most beneficial, is to maintain variation to select from. By crossing different varieties every year, they generate plants with new combinations of traits, ultimately new varieties to select from. For any crop, the sources of variation that breeders can use are the varieties grown around the world, along with their wild ancestors. Almost all the varieties of bread wheat that are currently grown are the result of crosses between other varieties grown somewhere else. Wheat's wild ancestors include three species: goatgrass (*Aegilops tauchii*), cultivated emmer, and if you go back further, two presumably extinct species—one a relative of einkorn and the other another goatgrass. But bread wheat's ancestors have different chromosome numbers; crossing a plant with 42 chromosomes with another that has either 14 or 28 chromosomes is almost never going to be successful.

The archetypical source of variation for wild species is spontaneously generated variation in the field. The likelihood of a random mutation being adaptive for food production is extremely low, and mutations are better for generating variation across an evolutionary time frame. Nevertheless, an initiative arose in the early twentieth century to apply new technology to do just that. X-ray machines were wheeled into the field in the 1920s. The idea was that random though it might be, the X-rays could bring about genetic changes, some of which might be beneficial. X-rays worked to generate variations in gladiolas, and experiments were conducted with other species, from rice to trees. Wheat was the beneficiary of an attempt to eliminate susceptibility to rust, one of the more devastating fungal pathogens. The logic was that if, as some assumed, a single gene conferred susceptibly to attack from the fungus, a healthy dose of X-rays might modify that gene, potentially reversing the susceptibility. It didn't work, and the practice of using X-rays to generate "hopeful" mutations eventually disappeared.

A second attempt to introduce new variation, this time using a chemical, was somewhat more targeted in that the chemical, colchicine, naturally occurs in the autumn crocus (a fall-blooming lily), and led to more specific effects than the application of X-rays or gamma radiation, which was also tried. This lily, *Colchicum autumnal*, had been used since ancient times for its medicinal properties, likely discovered because its active ingredients are strong enough to kill the sheep and cows that graze on it. But colchicine isn't all bad; it can be used to calm the arthritic symptoms of gout, although the medicine needs to be taken cautiously. The effect of colchicine which interested breeders is that it can trigger chromosome doubling, so the plants treated with it become polyploids. Doubling of chromosomes is of course part of the story of the wheat lineage. Each of the hybridizations, from einkorn to emmer and from emmer to bread wheat, was accompanied by a chromosome doubling. In fact, polyploidy has occurred in both plant and animal lineages as part of the evolutionary diversification of life-forms. As a breeding tool, it has been used more commonly on horticultural species than food species. A supermarigold, the Giant Tetra, was created by chromosome doubling caused by treatment with colchicine.

Our increasing capacity in chemistry and biology and the realization that science could be applied to help ensure a steady and secure source of food brought about another piece of the story of our relationship with our food species: agricultural research stations. The first stations in the United States were in Connecticut and California, established with private and state funding in the mid-nineteenth century. The introduction of commercial fertilizers was the impetus for these stations' creation. Fertilizers were increasingly available and heavily marketed, but they ranged from being good products to nothing more than snake oil. Farmers wanted field stations to run trials on fertilizers to determine their contents, and whether they increased plant growth in the field.

In 1862, a year into the Civil War, Congress established the United States Department of Agriculture (USDA). Much of its mission and more than half its resources were given over to a seed distribution program. Basically, it provided seeds to people interested in growing vegetables in their home gardens, and to farmers who could expand the area where commercial crops were grown. The early years of the USDA were a grave disappointment to the scientists who were hired, mostly from Europe, to conduct agricultural research, as they had little to no budget to support their work. Not surprisingly, the turnover among the researchers was high in the first decades of USDA history, although the seed distribution program expanded every year with encouragement from Congress, whose members were lobbied by their constituents who appreciated the free seeds.

In 1887, Congress passed the Hatch Act, establishing a mechanism to fund research in each US state and territory in support of developing a national agricultural industry. That structure remains to this day, and each state has multiple research stations. When breeders produce a new variety of wheat or another crop, they test it by growing it at stations located in different climate regions in their state and measuring crop yield across regions. Federal and state funding has ensured that agricultural innovation is publicly funded and accessible to farmers, and that the knowledge generated in research universities is transmitted to farmers. These provisions address a tension inherent in food production: like water and shelter, food is a requirement for life. How do we ensure that everyone has access to the food they need? The uneasy balance between government versus free-market control is expressed in almost every component of our food production systems.

The first states were organized around the need to provide food for people. With the advent of scientific research, we have made great strides toward uncovering the basic mechanisms underlying plant growth and crop productivity. Ideally, that knowledge could be put toward ending world hunger and increasing human health by improving nutrition. Unfortunately, our capacity to do good with our burgeoning knowledge is countered by our capacity to turn that knowledge into a tool to further our own gains at the expense of others. So the story of wheat continues.

CHAPTER SEVEN

War and Peace and Wheat

Our potential for both good and evil, as well as our capacity to muster our talents and skills to both nurture and kill, is a tale of humanity that we tell and retell. So it should come as no surprise that wheat, like many other foods, has been used as a weapon.

With scorched-earth policies in times of war, fields were destroyed either by the retreating inhabitants to leave nothing for their captors, or by the aggressors as a final insult after a win on the battlefield. As a staple food, wheat was especially vulnerable to being manipulated as agricultural systems matured. During the nineteenth and twentieth centuries, agriculture in Europe and North America shifted from small, diverse farms designed to support most of the food needs of a family and the nearby community to larger farms specializing in one or a few crops; nations depended on one another for imports to round out their diets.

The implications of this dependence became clear during World War I. It was no longer necessary to destroy fields where food would be grown; enemies could block incoming food supplies. The outcome of that war, a conflict between countries with comparable fighting power, was decided by hunger. Britain, after analyzing Germany's food sources, concluded that it could fight the war with food blockades and only a light military investment. Germany, considering the implications of the Allied troops' blockade of its food supplies, decided to strike hard in France to decrease the chance that the war would outlast its food reserve. But the war dragged on, and hunger was rampant across Europe. For countries on both sides of the conflict, there was a shortage of people to work the fields, as men were conscripted and war industries took resources, including people, from agriculture. And even after the 1918 Armistice, the Allied food blockade continued. Months passed as the Allies used the threat of starvation to pressure Germany into signing the Treaty of Versailles. That treaty was written without input from Germany, and required that the country pay reparations of $33 billion for damages incurred to civilians during the war. Before the Armistice, the German people without adequate food could blame their hunger on their own government's management of wartime food supplies; but after the Armistice, with the delay in removing the food blockades, their hunger was blamed on the Allies.[1]

While many of the wheat fields in Europe lay abandoned during the war, the opposite was true in Australia and North America. The United States had agreed to provide wheat for the Allies, but because of weather, 1915 and 1916 were low harvest years. International suppliers offered to pay more for wheat, causing prices in the United States to jump. Housewives protested when Chicago bakers doubled the price and halved the size of a loaf of bread in 1916. The National Housewives League teamed up with the National Association of Master Bakers to call for restrictions on wheat exports until domestic needs could be met. The government responded to the rising food cost and the calls for a wheat embargo in two ways. First, it encouraged Americans to eat less bread. Propaganda campaigns promoted two wheat-free days a week and one wheat-free meal a day as patriotic war efforts. By economizing on domestic wheat use, the United States could ship the surplus to Great Britain. Second, the government set a minimum price for grain at a level significantly higher than what farmers had been making, to encourage more wheat production. The price controls were controversial for both political and practical reasons. They were seen by some as a move toward socialism, a critique against a politician as potent then as today. Those who believed in a market economy had confidence that the market would stabilize food production without government intervention. Plus, given past experience, some people were concerned that merchants would hoard grains to further increase prices.[2] Moreover, farmers and millers were worried that setting a minimum price would also mean setting a ceiling on what they could earn. But ultimately, the strategy was very effective—the acres of wheat farmed in the United States increased from 47 million in 1913 to 74 million in 1919.[3]

Another initiative to increase wheat production for the war effort was focused on Indian lands, because many saw this territory as the easiest to access for a large-scale increase in farm acreage. One of the most outstanding examples of this happened in Montana, initially on two Indian reservations, the Fort Peck and the Crow, in the eastern part of the state. The initiative was a harbinger of the future of agriculture, and reflective of the effects of industrialized farming on Native peoples worldwide. Tom Campbell, an engineer originally from North Dakota, had faith in the economics and mechanics of large-operation farming. He negotiated an agreement with the Bureau of Indian Affairs (BIA) to lease up to 200,000 acres of tribal land for large-scale mechanized wheat farming.[4] Although Campbell's request—for 10-year leases, instead of the usual 5-year agreements, and for payments made in a percentage of the crop rather than in cash—was inconsistent with BIA regulations, government interest in fulfilling its agreements on wheat shipments to the Allies was enough to motivate flexibility with the rules. Campbell worked out the financing for the farming corporation with J. P. Morgan and three other bankers, who collectively put up $2 million to get it off the ground. The startup money funded labor and seeds and fences, but also equipment: 34 tractors with 35-horsepower motors (6-horsepower engines were common at that time), 10 threshing machines, 4 combines, 60 grain drills for seeding, 50 discs, 50 ten-bottom plows (10 plow blades are on one attachment), 10 trucks, and 100 grain cars (for transporting wheat from the field to the railroad). In its first years, the farm operated at a loss, in part because of the weather: 1919 was the driest year on record in that part of Montana. When the war was over and price controls for wheat were discontinued, corporate investors were eager to divest. Campbell purchased the equipment and the leases from the banks for a small fraction of the original cost, confident that as soon as the weather cooperated, he would make money.[5]

The Campbell Farming Corporation was the first large-scale mechanized wheat farm, and in many ways reflected the spirit of American capitalism in the twentieth century: invest large amounts of money, apply the newest technological developments, and make money. The number of farmworkers needed to run such farms was drastically reduced, an important consideration when this labor force was conscripted. It also meant that these workers could be paid on a level commensurate with what they could earn working in a factory, an equally important consideration for people trying to make a living. Campbell, wanting to underscore the efficiency of mechanization, set a record in

1924 using 15 of his tractors, each one with four implements attached: a plow to turn over the soil; a disc to break up the surface; a drill to seed the fields; and a packer to tamp down the soil after seeding. His men and equipment plowed and seeded a 640-acre field in 16 hours, which would have required 1,800 horses and 500 men to finish the task in same amount of time.

Mechanization changed the reality of farming so that food could be grown without using slaves, children, or unpaid workers. When Thomas Campbell started farming in Montana, 26% of the US population was employed in agriculture. Compare that to 90% during the medieval period in Europe, and 50% by the nineteenth century. Today in the United States, that portion of the population is closer to 1%.[6] As production systems grew in complexity and technological capacity, more people were able to work off the farm, in villages or cities, in markets or factories. The decline in the demand for agricultural workers is part of a recurring story over the next decades: the small farm, relying on unpaid family labor, was in no way competitive with the large corporate farms. Family farmers had to make the transition to spending money for mechanization and increasing acreage, or their farm went out of business.

After World War I, both Europe and the United States were struggling. Europe focused on rebuilding its cities and farms, leaving the United States with a surplus of grain. American farmers' incomes dropped, the beginning of the downward spiral into the Great Depression. The financial breakdown that led to the economic crisis was accompanied in the United States by an ecological crisis, starting in the wheat fields of the midsection of North America.

The grasslands that stretched from central Canada to Mexico had covered over half a million square miles (320 million acres) before Europeans arrived. Rainfall and temperatures vary across this region, so in the wetter northern areas, the grasses grew well over a person's head, but only about knee high in the drier sites. The southern plains, encompassing parts of Texas, Oklahoma, Kansas, Colorado, and New Mexico, are the driest, hottest, and windiest part of these lands. The soils there are held in place by the tenacious roots of the grasses and forbs that have learned to thrive amid the uncertain weather and the large grazers. Native Americans had used this land long before Europeans arrived, most recently as hunting grounds for the large herds of bison that started to decline in the nineteenth century and then were hunted to near extinction by agents of the US government, in another example of using food as a weapon.

The lands from the Mississippi River to the Rocky Mountains changed hands from the Spanish to the French and then in 1803 to the US government as part of the Louisiana Purchase. After bison were exterminated and Native peoples sent to reservations, the settlers' use of the southern plains alternated between cattle grazing and row crops. Ranchers overstocked the lands, reducing the quality of the range for cattle and leaving them vulnerable to droughts and hard winters. When the ranching industry crashed, homesteaders would move in, plowing up land to plant crops. Then when harvests were bad, the homesteaders didn't have a cushion to survive the slim years, and ranchers bought up the abandoned homesteads. But railroads, banks, and land speculators had financial incentives to increase settlement in the drier parts of the West, and with that they helped promote a dry-farming movement, proposed by a farmer from South Dakota. Hardy Webster Campbell (no relation to Thomas Campbell) argued that "the vast prairies of the semi-arid belt are not simply for the grazing of a few scattered herds, but for the support of vast numbers of smaller herds and flocks and thousands of ideal farm homes interspersed with numerous flourishing towns and cities."[7] *Progressive Agriculture: Tillage, Not Weather, Controls Yield*, the title of Campbell's later work, reflects his recommendations for dry farming, encouraging settlers to make the arid regions of the nation's midsection their home site. At that time, while Allied troops were fighting on European soil and the US government was offering a good price for wheat, homesteading the drier southern plains seemed like a real option. Homestead size increased to 160-acre lots, and farmers supported the war effort by plowing up every inch for wheat fields.[8]

The patriotic efforts to produce more food might look good on paper, but on the ground they were a disaster in the making. The soil is a farm's most valuable resource. Although the sunshine and rainfall are out of the farmer's control, the soil is not. It's not that farmers and government advisors ignored the soil; it's more that the methods promoted by dry farming hadn't been tested by agricultural scientists across sites and farming systems and, most important, in times of serious drought. The dry-farming methods encouraged packing down the subsoil, then using both plows and disc harrows to create a layer of dry soil on the surface that functioned like a mulch cover to retain water. The idea was that rainwater could be stored at depth, within the soil column. When soils dry and their surface cracks, air channels form down to the depths, enabling evaporation of water in the subsoil. By packing the soil, the air channels are compressed, and then with a dusty layer of soil on top, there's no direct path for water to evaporate. Thus, rain could be stored in deeper soils, where roots could access the water when needed. By applying Campbell's method, the soil surface was broken up after it rained, working between the rows when crops were young.

Although most farms bore little resemblance to the large-scale wheat farm of the Campbell Farming Corporation in eastern Montana, government price supports during the previous decades had bolstered at least some of the farmers enough to replace their horses with tractors. That meant that soil could be turned over at a faster pace, using both steel plow blades and the disc harrow, a beam with a row of 14-inch discs that, when pulled across the soil, could break the surface layers into small pieces. These methods may have been effective in some years, but the 1930s were dry across the whole United States, especially in the southern plains. Rainfall was less than half the normal average in areas already at the limit for supporting agriculture. Without spring rains, crops didn't grow, and the fields lay bare to bake in the sun. Those fields, with packed subsoils and loosened surface soils, were especially vulnerable to wind erosion.

The dust storms happened not once but many times a year, from the early 1930s to the early 1940s. Clouds of topsoil blackened the skies and moved across the landscape like dark blizzards. The storms lasted anywhere from one hour to three and a half days. Sometimes the dust was so thick, people couldn't see their hand in front of their face. Storms in 1935 destroyed 5 million acres of wheat across Kansas, Oklahoma, and Nebraska. Drifts of soil blocked roads, and people were instructed to stay inside. Inside what? The soil came into homes through the windows and the cracks in the walls, settling everywhere. Dust pneumonia, bronchitis, and respiratory diseases followed. In one storm alone, the wind picked up 350 million tons of soil, 12 million of which were deposited in Chicago, while the rest produced a haze across the sky in the East. Some areas of the Great Plains lost 75% of their topsoil. A million people moved out of the southern plains during the first half of the 1930s, and 2.5 million during the second half. Some moved because their fields were ruined, but more left because of financial hardship or panic. The American dream of plowing up land to make money growing wheat wasn't as easy as it seemed. It took more than the optimism of the banks and railroads and land companies to make a good harvest.

While winds were eroding the topsoil in the Midwest, countries on the other side of the Atlantic were struggling as well. The

Great Depression in the United States had ripple effects across Europe. Germany was straining under the burden of the reparations it owed from World War I, and an equally depressed economy.[9] During these times of limited food and growling stomachs, the nationalist rhetoric of Adolf Hitler and the Nazi Party appealed to people's fear and anger. Hitler was particularly adept at generating support at rallies, where he extolled his platform of doing away with politics, taking over the army, expelling "outsiders" from the country, and building a strong economy.

Hitler ran for president of the German Republic in 1932, but he didn't have enough support from the electorate for a win. In the same election, however, the Nazi Party gained the highest number of seats in the German parliament, though still only 30%. With a parliamentary form of government, the prime minister, or in Germany the chancellor, is selected from the party with the most members in the legislative body. After forming a coalition between the Communist and the Nazi Parties, Hitler became the chancellor. The powered elite thought that they could use Hitler—he was so effective at generating support from people around the country—to build an authoritarian and conservative government, and they were confident that he could easily be kept in check by more traditional politicians. They were right on the first count, and wrong on the second. In 1933 the Reichstag, the building that housed the parliament, burned down. Culpability was unclear. Hitler blamed it on members of the Communist Party; others believed that Hitler was responsible. Regardless, he used that incident to enact a clause of the German constitution to assume all powers in times of crisis. When Germany's president, Paul von Hindenberg, died the following year, Hitler took over the powers of the presidency and named himself both führer and chancellor.

The necessary first step for building the German economy was establishing a stronger agricultural base. That meant increasing the productivity of the republic's farms by consolidating smaller farms to increase overall efficiency. And it meant adding to the agricultural land base so that Germany could achieve food autonomy. Hitler wanted what European settlers in North America had—access to vast plains like those the settlers had taken from the Indians. The fields of Ukraine, known for their fertile soils and potential to grow wheat, were the answer. By consolidating its farms and then moving the displaced farmers to the east, Germany could take over food resources from Soviet republics and grow enough food to support both its army and its civilians. As with the western plains of North America, there were people living in the lands to the east of Germany, so their fate had to be considered. The Hunger Plan, developed by Nazi strategists, was based on the idea that the Soviet Union could be the resource base for an expanding Germany. That would work, of course, as long as none of the food grown in Soviet fields got sidetracked to feeding Soviet citizens. Put more bluntly, the plan called for the starvation of 23 million people.

Hitler's offensive began in 1939, after signing a Treaty of Nonaggression with the Soviet Union. The terms of the treaty stipulated that Germany would divide up Poland with the Soviet Union, and in exchange for a supply of food and fuel and raw materials from the Soviets for Germany, it would not attack the Soviet Union. With the invasion of Poland, Europe was drawn into war with Germany, and World War II began. By the end of 1940, Germany had taken over France, Belgium, Norway, Denmark, and Holland. Great Britain was a problem, with its powerful military and naval forces. And food was still a problem. The countries that Hitler had conquered all had struggling food economies. The war effort diverted fuel for tanks and fertilizers for explosives. None of the occupied countries were autonomous for their food needs, and many of them had relied on fertilizer or fodder from Great Britain. So Germany was again faced with the problem of how to feed its troops and its people back home. The obvious choice was to ignore the Nonaggression Pact and attack the Soviet Union before the United States would be drawn into the war.

Germany invaded Ukraine in 1941, taking over cities and villages and rural areas where many people were hungry and unhappy with Soviet rule. The collectivized farming imposed by the Soviets in the early 1930s meant that farmers lost their right to their own fields, people were required to work in the collectivized fields, and the harvests went into a common pool. In exchange, rural peoples were paid a nominal wage, barely enough to cover their food needs, and perhaps allowed time to work in their own gardens, where they raised chickens or cows and grew vegetables. Many peasants lived without the bare necessities; rags wrapped around their feet substituted for shoes, and houses were small huts. Life was so hard under Soviet communism that some Ukrainian villages welcomed the invading German soldiers and treated them like family.

The war continued for 4 more years. Its atrocities have been documented and described and memorialized to keep the history present enough so that we can deflect movements that dehumanize people based on their nationality or religion or color. The original Hunger Plan failed, partly because of the disorganization and conflict among different factions of the Nazi Party. Moreover, the logic of conquering agricultural lands to feed the German people, besides the horrific aspect of starving other people, missed the point that wheat fields don't grow by themselves. Plants had evolved long before humans and can grow without our help, but agricultural harvests aren't the same. When people are treated like expendable tools, used to achieve a regime's objectives, the essence of farming gets broken down into a simple working of the land. But starving workers with no hope of rectifying their hunger aren't efficient farmers. Food production across Soviet republics had been low before the German invasion, and productivity levels did not improve during the war. The failure of the Hunger Plan meant that Germany's soldiers were underfed, and its women and children back home were hungry. Only those in power ate well. The Nazi Hunger Plan was based on withholding food as an act of war, an approach used since at least the earliest documented warfare, though this time on a larger scale and in support of racist and nationalist ideologies. Our story of wheat—the wonder of being able to figure out how to use these grass seeds as a food, how to grow the plants, how to manage wheat fields across large geographic areas with different soils and different climates—has turned dark.[10]

The United States entered World War II in 1941, after the Japanese bombing of Pearl Harbor and Germany's declaration of war on the United States. For the wheat fields of North America, the war on the European continent meant that the United States and Canada had a market for all the wheat they could grow. Those wheat fields weren't supporting only American consumers but the American military, along with Britain, China, and the Soviet Union. War wasn't easy on farm production. Farm laborers were drafted, the government diverted fuel for military uses, factories that produced agricultural machinery switched production to tanks and arms, and the chemicals for making fertilizers were instead used for making explosives. The nation's transition to mechanized farming that was heralded by the Campbell Farming Corporation continued, although not at the same scale for farmers without financial backing from J. P. Morgan. Tractors replaced horses, and combines did the work of both reapers and thrashers. The shift to mechanized farming meant that farmers were now more limited by their ability to afford machinery than they were by labor.

With the United States, Canada, and Australia supplying the wheat for most of the war-torn countries of Europe, the agricultural sector was pulled out of the deep funk of the Great Depression; US farm incomes rose by 150% during World War II, as did the number of acres either newly plowed or plowed after having been abandoned. After the dust storms of the 1930s, federal programs encouraged farmers to take seriously eroded lands out of cultivation and to plant grasses for forage.[11] Alternatively, other advisors argued that better farming methods, with contours and terraces added to the fields to act as windbreaks, could support higher production levels in the southern plains. But while conservation farming methods were encouraged, nothing was regulated. As a result, when the demand for wheat increased during the war years, plows turned over the soils that had been ravaged by the dust storms. Wheat fields spread across the landscapes, and by luck and good fortune, the drought and the dust storms that came in the 1950s were less severe.

With food production increasing at a regular pace from the 1940s onward, US food exports became an important part of global food security. Instead of just thinking about food as a weapon to be withheld, Americans now had the opportunity to use it for humanitarian aid. The United States had made the shift from subsistence to commercial agriculture, based on the mechanization of farming and the use of fertilizers (and, increasingly, pesticides) to manage the challenge of growing large fields of monocultures (crops of single plant species). Even though we coddled our food crops by preparing fields to give plants the optimum conditions for growth, we also broke down one of nature's biggest defense systems: the ability to hide from predators. Instead of growing in scattered stands with a mix of other species, wheat was now being grown in large fields, with nothing but other wheat plants. Consequently, insects that burrow into stems or fungi that parasitize leaves had a big target: fields full of the aromatic smells that pathogens and herbivores use to find their favorite prey. Our close relationship with our crop species had a ripple effect on other species that have an interest in wheat. We engineered new environments that were optimal for predators of wheat, and in turn the predators of those predators.

Problems with insect pests were especially pronounced for North American farms because of their propensity for regional specialization in single crops to maximize production for an export-based economy. The first centuries of farming wheat in the United States read like a who's who of organisms that excelled at sharing in our bountiful harvests. First the rusts, a group of fungi that can decimate wheat fields as leaf rusts or stem rusts, black rusts or striped rusts. Then there are the smut fungi—variously named loose smut, stinking smut, flag smut—that infect the seed heads and release a myriad of tiny black spores, hence their name (*smut* is derived from the German word for "dirt," *schmutz*). Wheat scab or head blight is caused by yet another fungus, which leaves toxins in the grain, rendering the crop pretty much worthless. Moving from fungi to insects, the young maggots of the Hessian fly suck the sap from wheat plants; the grain midge, an orange mosquito-like fly which arrived from Canada in 1820, feeds on the developing grains; and chinch bugs, which are native to the United States, prefer corn to wheat, but will settle in wheat stands growing near corn. Grasshoppers entered North American wheat fields like a flying plague, leaving just the stems behind. And that's just part of the list.[12]

Since the early days of agriculture, farmers have managed pests with chemicals. In Egyptian times, it was oil applied to the leaves, or sometimes cinders, and eventually arsenic compounds. It wasn't until the eighteenth century that naturalists began a concentrated effort to learn more about insects, their life cycles, and the possibility of controlling their populations by targeting a specific point of their life cycle. These scientists soon realized that insects could be controlled with other insects, that the enemies of our enemies are our friends.

The installation of agricultural research stations throughout the United States was important for providing advice and experimentation on the control of insects that had negative impacts on agricultural production. A major emphasis of this effort was the exploration of biological controls, finding the predators and parasites of the major pests that would keep them in check. Since many of the new pathogens were in fact introduced from other countries, by being transported either with the seeds or with some of the produce shipped across the Atlantic, searching the native pests' countries could lead to the discovery of their natural predators. But with the development of insecticides during the world wars to manage lice outbreaks on soldiers, we developed our expertise in the large-scale application of pesticides to ensure predictable harvests.

Much of the effort behind breeding wheat varieties in the late nineteenth and early twentieth centuries was aimed at finding varieties that could thrive in the diverse climates of North America, and varieties that were resistant to the rusts, smuts, and insects that fed on wheat. But after World War II, munitions manufacture halted, and the resulting surplus of ammonium nitrate now could be used to fertilize fields at a lower price. With that came a new challenge for breeders: finding varieties of wheat that could take advantage of high levels of fertilizers in the soil. Seemingly, any plant that grows well would grow even better with more of what it needs. But for wheat, that isn't necessarily the case. In the presence of abundant nutrients in the soil, wheat produces an especially tall flowering stalk. The weight of the seeds can exceed what the long stalk can support, so it folds, with the seed heads pointed toward the ground—an outcome termed lodging. Because lodging limits the capacity to harvest the grains, heavy fertilizer application actually reduced the amount of grain harvested.

To solve this problem, our story takes us to Japan. Wheat had been introduced there from China by at least 200 BC, and farmed more commonly after the introduction of stone grinding mills in the seventh century AD. Japanese farmers had long produced wheat and rice in the same fields, planting wheat in the fall and intercropping rice with wheat in the early spring, two months before the wheat was harvested. They planted a dwarf variety of wheat, one that allocated more of its carbon to grain than to stem, both increasing wheat production and serving as a better companion plant for rice, with less competition for sunlight. Another advantage of this Japanese wheat was that when the soil was fertilized heavily, the genes that controlled the dwarf characteristic prevented the stems from growing tall and lodging. This wheat variety, Norin 10, was brought to the United States from Japan in 1946. It wasn't entirely new; it had been derived in Japan by crossing a Japanese dwarf wheat with a couple other varieties: one brought from Pennsylvania in 1892 and the other Turkey Red, the same variety the Mennonites had traveled with between Russia and Kansas.

Once in the United States, Norin 10 was picked up by several wheat breeders, one of whom was Orville Vogel. At Washington State University, Vogel crossed the dwarf wheat with winter wheat varieties that grew well in the cold winters and dry hills of eastern Washington. Likewise, in North Carolina it was crossed with soft, red winter wheat varieties. But its most famous lineage derived from the seeds Vogel sent to a colleague, Norman Borlaug, who was leading a Rockefeller Foundation project in Mexico.

This takes us back into politics, but this time, instead of using food to support soldiers or to starve enemies, world leaders were considering how food could be used to prevent war. After World War II, there was an understanding, articulated by the United Nations in 1945, that food and hunger, no matter where they occurred, were every country's responsibility. The United States stepped up to the plate with the Marshall Plan, which provided aid for European countries to rebuild their economies. In 1949 in his inaugural address, president Harry Truman articulated his goal of spreading US influence by transferring its expertise to less industrialized countries, thereby encouraging capitalism and discouraging communism. The underlying political impulse was a realization that military alliances and a strong army were as important to our national security as the economic growth of undeveloped countries.

The sharing of expertise was funded through a combination of government and private dollars, with the Ford Foundation and the Rockefeller Foundation at the forefront. The involvement of wheat in peace-keeping efforts was initiated by a trip to Mexico made a few years earlier by Henry Wallace, who was about to start his term as vice president under Franklin Roosevelt after having been Secretary of Agriculture for 7 years. Wallace was an unconventional person, but perhaps the perfect person to make the 1940 trip to Mexico, standing in for Roosevelt at the inauguration of the new Mexican president, Manuel Avila Camacho. Wallace drove to Mexico with his wife in their own car, stopping to talk with farmers along the way, and met with the outgoing president, Lázaro Cárdenas, to discuss how the United States could help Mexico. Given the timing of the trip and the anticipation of increasing involvement in the war in Europe, Wallace couldn't promise anything on behalf of the US government. But at his urging, the Rockefeller Foundation financed a Mexican-American venture to improve crops grown by Mexican farmers; educate these farmers about new varieties and techniques to increase food production; and establish agricultural science programs in Mexico. The end goal was that Mexico could be self-sufficient in its food production.

The main agricultural crops in Mexico at that time were corn, wheat, and beans, and Norman Borlaug was hired to lead the part of the program that focused on increasing the production of wheat. A rust outbreak had reduced the Mexican harvest by half for 3 years in a row, so the first focus for Borlaug was to develop varieties resistant to stem rust. Borlaug was a Midwesterner, raised on a farm in Iowa and educated at the University of Minnesota. He studied forestry as an undergraduate; then plant pathology, researching the effects of Fusarium fungi on box elder trees for his master of science degree; and then, for his doctoral research, the effects of another species of Fusarium on flax. Borlaug had been recently hired at DuPont Chemicals when his former professor invited him to join the Rockefeller program.

The development of rust-resistant wheat was painstaking labor. Starting with a collection of wheat varieties from around the world, Borlaug selected three types: varieties that grew well in Mexico with high yield; varieties that matured early, to make efficient use of irrigation water; and varieties that were known to be rust resistant. Using combinations of parents with different traits, he began with over 200 crosses. The seeds produced by those crosses were then grown in test plots, and if the plants didn't succumb to rust, their seeds were saved and crossed again with the best-yielding wheats. Borlaug understood that following the protocols commonly used in breeding programs, it would take 10 years of breeding to isolate rust-resistant varieties. Impatient with that prospect, he devised a way to grow two generations of wheat each year: in the winter, he planted wheat in mid-November, the common planting time in Mexico, at sites in the Yaqui Valley in Sonora, near sea level. The seeds of the best-growing plants were harvested in early May, then driven up to the second, high-elevation field site, 8,000 feet above sea level near Toluca, and planted by early June at the latest. Summer was a time of frequent rains at the Toluca site, and the wet weather combined with warm summer temperatures were perfect conditions for rust outbreaks. Borlaug's approach of working at two different sites was controversial, running headlong into standard practices: plant breeding was thought to be optimal when done in the soil and the climate where the crop was to be grown. The sites in the northern Mexico valley and the high-elevation site differed in so many ways—short winter days versus long summer days, and different soil types, temperature range, precipitation, and pests. To many, it seemed like an approach doomed to fail.

Not only that, but crossing wheat plants is meticulous work. The flowers of the grass family are small, even tiny. Usually, wheat will self-pollinate: the flowers include both the anthers, which carry the pollen, and the stigma, which is the receptive surface on which the pollen lands and then germinates, producing a tube that fertilizes the ovum. If you want to cross two plants, you must prevent self-fertilization by opening the tiny scales of the floret, and with a pair of fine tweezers plucking the three anthers without losing any of the pollen in the process. Then in a day or two, when the stigma is mature (it will resemble a plume), you add pollen to the plant by carefully shaking the anthers from the plant you want to be the other parent over the recipient plant's stigma, bagging the flowering head to prevent any other pollen from entering, and hoping that the cross worked. Though Borlaug may have been impatient for results, he and his workers had to have had infinite patience to complete the exacting tasks associated with crossing thousands of plants twice a year.

Their efforts paid off. By 1948, Borlaug had seeds from four varieties of wheat that were resistant to rust fungal attacks and that matured early. As a bonus, because he had bred them at two different sites, those varieties were broadly adapted to grow well with either short or long days, in different soil types, and with contrasting climates. Having met the wheat program's first goal of generating rust-resistant varieties, Borlaug shifted his attention to maximizing harvests. Fertilizer and irrigation clearly increased wheat yields, so he set up plots at the experiment station with new and old varieties, with and without irrigation, and with and without fertilizer. The purpose was to demonstrate to Mexican farmers the advantages of more intensive agriculture on crop yield. But with irrigation and fertilizer, the problem of lodging, when tall stems bend from the weight of the grains, was apparent. Borlaug searched through the USDA collections of wheat for short varieties without much success. Then, having heard about semidwarf varieties brought from Japan, he contacted Orville Vogel, who sent seeds from the offspring of the semidwarf Norin 10 crossed with two varieties of Washington winter wheat. It took a few iterations before the crosses were successful, followed by multiple generations of breeding. But 8 years later, Borlaug had wheat varieties that were rust resistant and high yielding under irrigation and fertilizer.

The Rockefeller Foundation program was a success, not just for supporting the breeding program that generated high-yielding varieties of wheat, but also in its aim to build the science of agriculture in Mexico. The International Center for Wheat and Maize Improvement (CIMMYT) opened in 1966, a transformation of the original Rockefeller-funded office. The center, which partners with farmers, nongovernmental organizations, development organizations, and research institutes, has offices in 15 countries and projects in 40. Its germplasm bank includes more than 140,000 unique collections of wheat; more than 70% of the wheat grown in developing countries is from a pedigree that links back to a CIMMYT variety. The emphasis on training remains strong; over 10,000 students or scientists have trained at CIMMYT. The outcome of the Rockefeller project has been called the Green Revolution, in part because of the wheat yields in Mexico: with the use of rust-resistant and high-yielding varieties in conjunction with fertilization and irrigation, wheat production increased, roughly doubling the yields across time.[13]

While plant breeders generally aren't celebrities for having produced a successful cultivar that contributes to the food on our dinner table, Borlaug gained more attention from the public than all other breeders combined. In 1970, he was awarded the Nobel Peace Prize, in 1977 the Presidential Medal of Freedom, and in 2006 a Congressional Gold Medal. His work on high-yielding varieties that can take advantage of fertilizer inputs has earned him both accolades and curses.

While the varieties that Borlaug developed tripled or more the amount of grain that could be produced per acre, in Mexico they also represented a shift from subsistence agriculture managed by peasant farmers to commercial agriculture commandeered by farmers with enough capital to purchase fertilizers and install irrigation systems. For many, the Green Revolution isn't a tale of successful breeding but rather a story of capital-intensive agriculture that disrupted the way of life for many poor people. Subsistence farmers were displaced, and the number of people moving to Mexico City swelled, as did the numbers crossing the Rio Grande to come to the United States.

The Green Revolution reflected an approach to humanitarian aid that has its roots in US political theory of the early twentieth century. With the development of agricultural technologies and especially with the food surpluses that American farmers produced for decades after World War II, we understood that hunger and poverty were not necessary human conditions. Since the first states, like those of the Sumerians, the production and distribution of food have been controlled by the state. While the state imposed demands on its citizens, provisioning food was one of the advantages to living in one. Through time, with the rise of kingdoms, fiefdoms, nation-states, and world wars, it became clear to American policy makers that our assistance to other countries was important for humanitarian reasons, but equally important for maintaining the distribution of power in the world. The Green Revolution, in fact, was designed to counter the Red Revolution, the spread of communism. Hungry people, the logic went, were susceptible to communist ideologies. Exporting our scientific and technological expertise was a way to spread democratic and progressive ideals and help with the process of nation-building.

But the introduction of high-yielding varieties of wheat was pointless if not backed by government programs to encourage farmers to grow wheat, and financial assistance to either import or manufacture fertilizers. Agricultural technologies had huge effects on the development of societies. In the fledgling United States, with lands that we took from Native Americans, we had fields aplenty, but our farms were chronically short of labor. The shift to mechanization was expensive, and resulted in difficult transitions for much of rural America. The loss of the family farm and the deterioration of rural villages were painful for many people. But at the same time, it freed up people to pursue work away from the farm. And in a country with a thriving economy, children often left the farm for work in the city; some were attracted to the lifestyle change and the potential for making more money. For countries without a thriving economy, the shift to mechanization and expensive fertilizers and irrigation systems meant that many poor farmers lost their means of subsistence. Instead of finding opportunity in the city, their displacement could mean moving to urban slums with little promise of work and a livelihood. Although the economic gains of high-production agriculture are indisputable, the distribution of those gains has been a sore point—particularly from the perspective of the displaced.

CHAPTER EIGHT

Order in Chaos

The pressure on wheat to produce enough food for all the people and some of the animals is mounting. In 2019, 538 million acres of land on our planet were planted in wheat. That area is equivalent to planting every square inch of California—the Hollywood Hills, the peak of Mount Whitney, and the whole I-5 corridor—along with Washington, Oregon, Idaho, Utah, Arizona, Wyoming, and a little bit of Montana. The rivers, the lakes, the mountains, the freeways, the whole thing would be covered in wheat. The equivalent in the Midwest would be to plant the Dakotas, Nebraska, Kansas, Minnesota, Iowa, Missouri, Wisconsin, Illinois, Michigan, and Ohio. That amount of wheat would be enough to feed our 7.7 billion people. By 2050, our population is expected to exceed 9 billion. The demand for space for all those people will be in direct conflict with the need for more agricultural lands. That's a problem for species besides just humans and wheat, and it's exacerbated, because food demands increase faster than population growth. That's partly because we're eating more, but mostly because people are eating higher on the food chain—more meat—which requires more grain to feed the animals we butcher. With that, demand for grains could increase more than 40% by 2050.

How will we grow that much more food? The first Green Revolution resulted in a doubling of wheat production in the countries where the new high-yielding wheat seeds were planted, fertilized, and irrigated. The breakthrough in breeding for

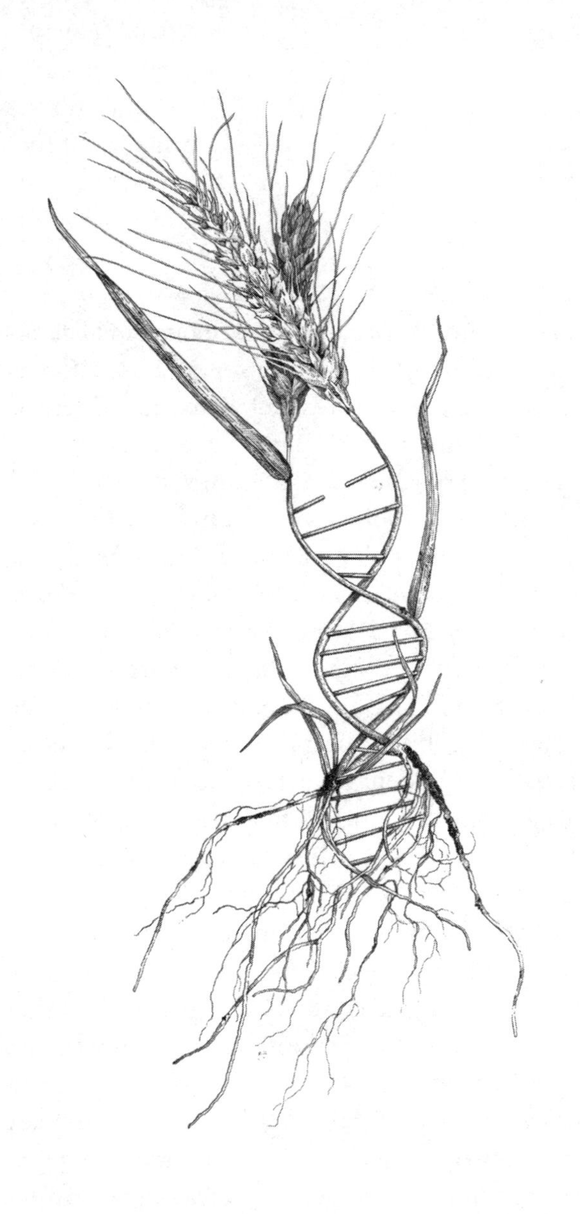

increased yields began with wheat, and the same breeding logic was then applied to rice, with similarly spectacular increases in production levels. Over the last decades, the technology was used on some nongrain crops that are staple foods for other countries. While the Green Revolution was important for increasing global food supply, regions without water for irrigation didn't realize the same yield increases.[1] But overall, the increased crop production meant the threat of famine due to low food supplies was largely eliminated. People still go hungry, but the cause is not because we haven't produced enough food globally.

With the gains from the Green Revolution, and a looming question of how to meet pending food needs, it's not surprising that people talk of a second Green Revolution, GR 2.0.[2] But what would GR 2.0 look like? The most common answer to that question is a hopeful nod to molecular technologies that, some say, hold the promise of catapulting breeding programs to a new level. And then there's the other side of the coin: having corporations making decisions about what food is available to us has loomed large in protest movements, with concerns that organizations answering first to their stakeholders don't necessarily hold the best interests of humanity as their primary motivation. And that brings us to the underlying question: will biotechnology lead to increased food production, initiating the next Green Revolution?

We finished sequencing the human genome in 2003; the wheat genome was mostly sequenced in 2018. Both projects were the culmination of years of work by large teams with vast amounts of funding. The human genome project carried with it the promise that once we understood the code of life, we could cure genetic diseases. It was funded to the tune of $2.7 billion, and though we have yet to realize a cure for most diseases, the funding supported the development of techniques that made other sequencing projects feasible. With that, the list of species with sequenced genomes grows each year. The progress made on sequencing the wheat genome is the result of a consortium of teams from 73 research institutes in 20 countries. Two hundred people coauthored the 2018 publication that describes the structure of each of the 21 chromosomes. In upcoming years, those same groups and others will finish the description of each of the identified genes, and assign locations to some segments of DNA that have yet to be mapped on the chromosomes.[3]

Sequencing a genome entails describing the order of each of the base pairs (adenine, guanine, cytosine, and thymine) on each of the chromosomes. It takes so long to sequence a genome because there are millions, tens of millions, hundreds of millions of base pairs on a strand of DNA from a single chromosome. To make the strand more manageable, we duplicate it many times, and then cut the strands at random points into shorter segments. Tracking the A,C,G,T's of short segments is pretty fast, and we rely on computer algorithms to find overlapping patterns between the fragments so that they can be aligned into contiguous segments (contigs, in biotech-speak). Scaffolds are then constructed with contigs, where the sequence has been determined with a high degree of certainty, and gaps, where the sequence is not clear. Scaffolds are then aligned to make up the structure of a chromosome. Then that sequence—millions, or tens of millions or hundreds of millions of base-pairs long—gets scanned to look for possible genes.

Possible genes are identifiable by looking for the same sequence that defines genes described in other species. We probably share about a quarter of our genes with wheat; corn and rice, whose genomes are already sequenced, have a much larger percentage of genes that overlap with those of wheat. The process of finding genes gets faster as our database of sequenced genomes grows. For genes that aren't shared with other species, there's a set of rules followed by all DNA and transcribed RNA: base pairs are translated into amino acids in groups of three. TGG codes for tryptophan, GAT codes for cysteine, GTT codes for valine, and so on. Sixty-one of the 64 possible triplets code for one of 20 amino acids, and amino acids line up to make proteins. (The other 3 triplets signal a stop for the cellular machinery that translates the RNA into amino acids.) Proteins direct cellular activity, things like duplicating and repairing DNA, transcription and translation from DNA into amino acids, and catalyzing the metabolic reactions that run the cell, from building carbohydrates during photosynthesis to breaking down ATP to release energy. All genes start with a specific 3-base-pair sequence (ATG) and end with one of three other 3-base-pair sequences (TAA, TGA, TAG). And in advance of every gene are promoter regions that guide the start of transcription. With that information, you can bookend the sequences that are potential genes, so it's pretty straightforward to write the computer algorithm to identify a possible gene.

What makes it a possible gene instead of a gene is that what could be an ATG start sequence might also be the code for cysteine followed by valine: G**AT G**TT. If you read the sequence starting one base to the left or right from where you should, it throws all the codons off. The other complicating factor is that in most genes, there are DNA segments in the middle of the gene that aren't part of the protein-coding DNA. These segments, sometimes short, sometimes long, get transcribed into RNA, but then are trimmed out when the cellular machinery gets the RNA ready for translation into a protein. What these noncoding segments are doing there and where they come from isn't clear.

Even with the challenge of not knowing where to start reading the sequence (there are only three possibilities), why should it take 15 years to sequence a genome? Computers have ever-increasing processing capacity and speed, after all. It's because finding the possible genes is the easy part. Our genome and wheat's genome are mostly not genes. We have 46 chromosomes;

bread wheat has 42. Our 46 chromosomes carry about 20,000 genes; the 42 wheat chromosomes may include over 100,000 genes. Of all the DNA that we carry, just a hair over 1% of it is genes; 98% is DNA that doesn't code for proteins. In wheat, the proportion of genes to total DNA is similar.

In the late 1960s, scientists learned that much of our DNA is noncoding—that there are a lot of copies of DNA composed of highly repetitive sequences. At the time, that DNA was dismissed as "junk DNA" or "selfish genes," pieces of DNA with no use to us but with the capability of being in the right place to get replicated and passed on. But we each have trillions of cells, some of which, like our skin cells, replicate every few weeks. For our bodies and for wheat plants to be copying that much DNA that doesn't code for genes, there must be some value to it, particularly since our cells have mechanisms to rid themselves of unnecessary DNA.

Noncoding DNA challenged our idea of genomes, originally conceived as the material that replicates and encodes proteins. Before sequencing the human genome, scientists debated whether they should sequence all the DNA or just focus on what they knew were the protein-coding sections. The argument for sequencing only the genes was based on expediency: given the volume of highly repetitive DNA and the likelihood that sequencing technology would soon become faster and less expensive, scientists could focus on the extraneous DNA later. Fortunately, the decision was to move forward with the entire sequence. While we are still not clear on the function of much of it, scientists can at least classify the DNA based on its structure, imposing some order in the face of what looks like chaos. So the non-protein-coding DNA gets categorized into short or long pieces (length determined by when it starts repeating the same sequence), pieces with repeating sequences at the end (terminal repeats), or groups distinct enough to merit a specific family name like Gypsy, Mariner, Harbinger, and Mutator. And what is left over is labeled Unclassified.

About half the human genome and 85% of the wheat genome are made of or derived from a special kind of DNA, familiarly known as jumping genes and more formally known as transposable elements or transposons. The jumping genes moniker describes their capacity to move to a new spot in the chromosome or to a different chromosome altogether. Barbara McClintock was the first to identify this phenomenon in corn chromosomes. Corn lends itself well to genetic studies, because each kernel on a cob represents the cross of two parents. The tassels at the top of the plant produce the pollen, while the silk is the receptive surface where pollen lands to grow a tube that can fertilize the ovum, ultimately producing one seed, or kernel. If you add the pollen from a single plant to the silks of one developing ear, up to 800 kernels develop. A comparable scenario in humans would be if you wanted to study variation between siblings, and you found a family with 800 children.

Scientists had established that one gene determines whether the corn kernels are the typical white to yellow color, which is the dominant trait, or conversely whether the kernels are darkly colored (the recessive trait). If both copies of the gene (one from each parent) have the recessive version of the gene, a second gene determines whether the kernels are purple or brown. While Gregor Mendel studied the large patterns of uniform response after crossing specific pea plants, McClintock's interest was in those instances when crosses didn't result in the expected patterns. Some of the kernels, instead of solid yellow or purple or red, had streaks or splotches of a contrasting color.

McClintock combined studies of inheritance patterns in kernel color with microscope work documenting the structure of the chromosomes. DNA usually occurs in long strands in the nucleus of cells. But when cells make copies and divide, the DNA condenses to form the X-shaped chromosomes. It does so with the help of proteins that act like spools—the helical strand of DNA wraps twice around the protein. The coiled strands then zigzag back on each other and fold in a very specific and predictable way, forming the chromosomes. With special staining techniques, McClintock recorded differences in chromosomes, down to a little knob in the top part of the short arm of chromosome 9. For corn with a streaked kernel pattern, that knob disappeared. That had to have been caused by a break in the DNA. McClintock posited that a piece of DNA (the transposon or jumping gene) could literally cut itself out of the chromosome at one site and insert itself somewhere else. In this case, the insertion changed how the DNA folded, causing the disappearance of the knob from the top of the ninth chromosome. Not only did the transposon change the surface structure of the chromosome, it also disrupted the function of the gene that determines kernel color, by landing in the middle of that gene. The malfunction of the color gene caused the stripes and splotches. McClintock could literally see the transposon that caused the variegated color patterns. To confirm her findings, she designed crosses between corn plants to verify that in fact, movable elements within the DNA were affecting both chromosome break patterns and kernel color.

McClintock was a respected scientist, but her colleagues were incredulous that genes would move around within and between chromosomes. At the same time, they couldn't find an alternative explanation for the results of her carefully designed experiments, and so jumping genes were considered an anomaly unique to the corn genome. Ten years later, scientists learned that some viral DNA functioned in much the same way, by breaking the chromosome and then patching it up to include the genetic material from the virus. Then the same phenomenon was discovered in bacteria and subsequently in other higher organisms. The scientific community finally understood McClintock's work. In 1983, 30 years after she first presented her results, she was awarded a Nobel Prize for her discovery of transposons.[4]

Now we understand that transposons include in their sequence the code for an enzyme that cuts the DNA specifically at the start of the transposon, and pastes it somewhere else; in another kind of transposon, instead of cutting the transposon and moving it, the transposon stays in place, but copies of the transposon are pasted in new places. Where those copies get inserted can make a difference; inserted in the middle of a functional gene, it can interfere with the production of whatever protein that gene codes for. Or if the transposon lands in the promoter section that controls when and how often a gene gets transcribed, the gene can overproduce or underproduce the protein it codes for. The peppered moth exemplifies the effects of transposons landing in the middle of a gene-coding sequence. The dark variant of the moth flourished during the Industrial Revolution, when factory soot coated the bark of trees. Peppered moths got named for their white-and-black coloration, but the dark variant is the result of a transposable element inserted in the middle of a coding section of a gene responsible for wing development. The presence of the element results in the overproduction of dark-colored wing cells, a variant that made a difference for the moths that blended into the background when resting on the soot-covered birch trees, thereby escaping predation by a hungry bird.

Think about this for a moment. Transposons can either move or have copies made and inserted somewhere else in the genome. That capacity puts pressure on the organization of the genome—especially when you consider that 85% of the DNA in wheat cells is either transposons or descended from transposons. The DNA in our cells contributes to our height, our skin color, our hair texture and whether we even have hair, our nose outline, our capacity to compose music or analyze data. For wheat, DNA affects the timing of seed germination, the number of leaves, the extension of the roots, the production of the embryo for next year's growth, and everything else that matters to its survival. If 85% of the wheat genome can make copies of itself and disrupt the functioning of all the rest of it, how, you might reasonably wonder, do we grow fields of wheat enough to mill and package and stir and bake into our daily loaves?

This isn't a perfect analogy, but consider Hartsfield-Jackson Atlanta International Airport, the busiest airport in the world. Having 85% of the genome composed of transposable elements is like asking a kindergarten class to take over the air traffic control tower at Atlanta Airport . . . maybe not take it over entirely, but for every trained air traffic controller, we add a busload of kindergartners to the mix. For the same reason that Atlanta Airport couldn't function with busloads of 5-year-olds in the control room, organisms have figured out how to keep transposons from jumping around.[5]

Some of the regulation involves the same mechanism that allows cells in the leaf to function differently from cells in the roots. Way back in time, when the first multicellular plants evolved and the cells differentiated into either body cells or reproductive cells, the plant had to silence the genes necessary for reproduction in the body cells that didn't reproduce, and vice versa. The DNA can bundle itself into a tight wad that makes it physically inaccessible to the enzymes that transcribe DNA into RNA. That strategy doesn't work, however, if there are other genes nearby that need to be transcribed. Another way to regulate whether a piece of DNA gets transcribed is through transcription factors, proteins that bind to promoter regions and attract the cellular machinery that starts transcribing DNA. And in a beautiful ouroboros—the Greek image of a serpent eating its own tail—transposable elements are hypothesized to be the catalyst and source material for the evolution of regulatory networks within cells, including transcription factors.

Gene regulation is what enables a multicelled organism to have differentiated tissues, and metabolic processes to vary at different life stages. That regulation can also happen through RNA molecules. Some of the RNA in the cell functions as the intermediary between genes and proteins. But as scientists are faced with trying to figure out what all the non-protein-coding DNA is doing, they have realized that RNA is a more versatile part of the cell than what was understood earlier. There are long, noncoding RNAs, microRNAs, hairpin RNAs (single-stranded RNAs that loop back on themselves to form lollipop-shaped structures), short interfering RNAs, and more. To keep the transposons under control, short strands of RNA attach to the DNA of the transposon, attracting proteins that act like a layer of duct tape on top of the DNA, keeping it from being transcribed or from jumping from one place to another. Other types of RNA guide a small molecule called methyl, made of a single carbon and three hydrogen atoms, to a point at the start of the transposon, preventing transcription of the DNA. The list of how cells regulate the transcription of protein-coding and noncoding DNA will no doubt keep expanding in coming years as scientists gain a better understanding of what we can no longer call junk.

Even the transposon, which I have unfairly equated with an unruly child, has potential utility. Transposons comprise 85% of the wheat genome, and about the same proportion of the corn genome. Besides thinking of them as a virus, something that infects the plant and replicates itself, some scientists have recently begun to investigate their evolutionary potential. When transposons move to a new part of the genome, they sometimes move not only themselves but pieces of the adjacent DNA—part of a gene, for example. Two nearby transposons sometimes move in tandem, and take along any DNA that separates the two. The DNA from the transposons, in concert with the fragments of other DNA that move along with it, can serve as the template with which to build new genes or perhaps a new way to regulate existing genes. This happens in particular during times of stress: drought or heat stress, or attack by a fungus or an insect, or a bath of herbicide with toxic chemicals. Under stress conditions, the control structures that keep the transposons in place are loosened, leaving them to do what they do best: possibly disrupting gene function, possibly increasing or decreasing the output of some genes. Some scientists argue that transposons, by changing the function of genes or being source material for new genes, are a built-in mechanism for organisms to generate the variation necessary for selection and adaptation to changing environments.[6]

To go back to the 1% that codes for genes, having the sequence of all the genes provides us with a database with which we can address questions about evolution. The evolution of plants, from single-celled algae to mosses and ferns and trees and then flowering plants, was the result of changes in genes and how they are regulated. When genes are duplicated, the second and third copies of those genes aren't under the same tight control as the original. Generally, organisms need their genes to function, so specific proteins read the DNA and correct any mistakes that might have been made when the DNA was duplicated. Like the youngest children of authoritarian parents, that quality control is relaxed for the later copies. Initially, the copies may serve to make more of the same protein, but over time the sequence can change slightly. It could happen through copying errors or a substitution or deletion of one of the base pairs. Or maybe a segment of the DNA gets inverted during copying, or perhaps a transposon lands in the middle of a copy. The changes can be enough to generate a slightly different protein. Those modified copies and the original make a gene family—groups of genes with enough similarity in their nucleotide sequences to recognize them as derived from one another, but with enough small changes to function differently. Gene families are common; some are small, with as few as two members, and others are large, with thousands of variants on the original.

Gene families are an important part of the story of where

wheat came from, how plants over hundreds of millions of years moved from single-celled algae floating in a pond to grasses and orchids and trees thriving on land. One such part of the story focuses on the cell wall, which provides the structural support needed in the plant's move from a buoyant environment to land. Its primary component is really simple—cellulose, a long chain of glucose molecules. But then those long strands link together to form microfibers, which are deposited into a matrix of other carbohydrates and proteins. The genes that code for the cellulose and the other carbohydrates of the cell wall belong to two gene families: the cellulose synthase gene family and the cellulose synthase–like gene family. Cellulose synthase manages the construction of the cellulose, adding one glucose molecule after another to make a long chain. With slight deviations in the DNA of the gene that codes for cellulose synthase, a family of similar genes is generated: the cellulose synthase–like family. And members of that gene family produce a palette of carbohydrates that form much of the cell wall. Then with further deviations, plants produce some cells that lay down a second wall inside the first, producing cells that are even stronger. Those cells, stacked one on the other, provide the internal structural support for the stem and leaves and trunks of plants that grow taller and taller. And they provide a system to move water from the roots to the highest leaves. With the sequencing of plant genomes, we can identify gene families that are common across many species; uncover the variants within those two gene families; and in time, possibly understand how the evolution of variation in those genes contributed to the breathtaking diversity of plant forms.

The picture that emerges—and will continue to emerge over the next decades as scientists add to our understanding of gene regulation and evolution—is that the wheat genome is big and messy. It includes three complete sets of chromosomes from three different parent species. And while those chromosome sets carry more genes than our cells carry, the chromosomes are primarily composed of DNA that doesn't code for proteins, though they may have a large role in regulating how all the DNA gets expressed. By comparing the sequence of genes within a single species and across plants that range from single-celled algae to complex flowering plants, we understand that evolution occurs through the copying and modification of genes and by differential regulation of those genes. A lot of the changes start out as really no more than a piece of DNA's random jump to a new spot, maybe changing how a gene is transcribed or what protein it codes for. Those changes, sometimes big and sometimes small, either work well or they don't. When they work well—say, the protein is modified to add to the complexity of carbohydrates in the cell wall—the plant has in some way changed how it interacts with its environment. And the evolutionary sieve of environmental factors that impact an organism between birth and reproduction and death will either allow the new variant to pass or stop it in its tracks.

Genome sequencing has shown us that we understand only a part of the code needed to translate the Book of Life. Our original dismay upon discovering that only a tiny percentage of the DNA in our chromosomes codes for genes is slowly being replaced with an understanding that the genome is a densely formatted database; that RNA is a messenger molecule and a regulatory molecule; that the repetition in the DNA may indicate that regulatory components need to be peppered throughout the genome to control the nearby genes; and that the evolution of the genome is at least in part driven from within, by changes generated by transposons reacting to cellular conditions. With all that going on, it's amazing that we can grow fields of wheat that are harvested and ground and baked into our daily bread.

Wheat breeding, as practiced in universities and research stations across the country, typically involves crossing a highly productive wheat variety with a second variety that has desirable traits. In the early decades of plant breeding, that process began by traveling afar to find varieties that grow in colder or drier or hotter or wetter climates. By bringing seeds from those varieties to new locations, we expanded the areas where we could profitably grow wheat, including the Great Plains of North America. Norman Borlaug developed varieties of rust-resistant wheat by using the principle that the most effective disease resistance includes multiple genes to counter different strains of the fungus. Then to increase the yield potential of the rust-resistant varieties, Borlaug searched for semidwarf wheats that reduced stem elongation. The semidwarf varieties from Washington State were crushed by the rust infection in Mexican fields, and it took some finessing to keep plants alive long enough to cross them with the rust-resistant Mexican plants. The seeds from that cross were grown and backcrossed repeatedly with the resistant plants from Mexico, until all the rust-resistant parent traits were present in the offspring but with the dwarf genes added. Because a cross produces an offspring with half the genes from each parent, each backcross results in a plant with incrementally more of the original, in this case rust-resistant, variety.

The promise of genetic engineering includes the transfer of just the relevant genes instead of a long series of backcrosses. Theoretically, Borlaug could have just added a single dwarf gene to his rust-resistant varieties, as opposed to making crosses and sorting through thousands of offspring for the ones containing the right combination of rust resistance genes and dwarf genes. It was decades too early for the technology, of course, and transferring just a single dwarf gene may not have had the same positive effect. Plants resulting from crosses with the semidwarf variety Norin 10 featured not just shorter stems but also more flowering heads and bigger grains; a larger proportion of each wheat plant was seeds instead of stems and leaves, translating to larger yields. Either the gene for dwarfing has effects on multiple traits, or multiple genes from Norin 10 were passed on.[7]

Genetic engineering, as controversial as it can be in both the United States and abroad, is an incontrovertible fact of our agricultural systems for some of our crops. In the United States, 94% of the soybeans grown, 91% of the cotton, and 90% of the corn are genetically engineered to be herbicide tolerant. Worldwide, the percentage of genetically engineered (GE) crops is still impressive: 77% of the soybeans, 80% of the cotton, and 32% of the maize.[8]

Most GE crops have been modified for resistance to herbicides, to protect the crop from herbivores, or a combination of both. Herbicide resistance, most commonly to glyphosate (trade name Roundup; the GE crops are called Roundup Ready) or one of several other herbicides, has simplified farming practices by allowing farmers to spray their fields in the middle of the growing season with an herbicide that kills all the weeds but doesn't affect the crop. Insect resistance is conferred to crops by adding a gene, isolated from soil bacteria, that codes for proteins (called Bt, after the bacteria *Bacillus thuringiensis*) that are toxic to some insect larvae, but break down under acidic conditions, such as in our stomachs and those of other mammals and birds and many insects. Because GE crops have fewer pest problems than their nonengineered relatives, they produce higher yields and, averaged across farms, make more money for farmers, even after taking the higher seed prices into account. Additionally, the crops with the Bt insecticide built into their genome require fewer pesticide applications, decreasing the use of those chemicals by an average of 9% globally.[9]

The downsides to the GE crop varieties hark back to the potential for life to evolve, for biological systems to generate variation in the presence of stress. Just as Borlaug knew to include multiple resistance genes in his rust-resistant varieties of wheat, a single line of resistance is a sure way for enemies to find a weak spot to enable them to overcome that resistance. Since the introduction of glyphosate-resistant varieties of soybeans and corn in the late 1990s, the use of glyphosate has increased fifteenfold. That increase isn't entirely due to the herbicide-resistant crops; no-till agriculture was widely adopted during that period. Tillage kills weeds without chemicals; it turns the soil over, burying the weeds. Early in the twentieth century, H. W. Campbell recommended a light tilling of the soil after every rain to create a dust mulch that would prevent water from evaporating from the soil. While that might have been true on some soils, tillage disturbs the soils, leaving them vulnerable to being blown or washed away. With the adoption of no-till agriculture, the plant material left after crops have been harvested remains on the surface of the soil, and that mulch helps the soil retain moisture while reducing wind and water erosion. But to control weeds, many no-till farmers have relied on glyphosate.

The increased and consistent use of glyphosate as a result of no-till and Roundup Ready crops resulted in a phenomenon that's no surprise to evolutionary ecologists—weeds became resistant to glyphosate. By 2019, 47 weed species had developed resistance to Roundup, and many of those varieties originated in fields growing Roundup Ready crops.[10] Glyphosate kills plants because it gets absorbed by their leaves and transported throughout the plant. When it reaches growing tissues, glyphosate molecules bind with a plant enzyme that's necessary for building a couple of key amino acids. Without those amino acids, the plant can't make the proteins it needs, so it stops growing and dies. The species that have become resistant to glyphosate's effects have outsmarted the herbicide in different ways. Some plants, instead of transporting the chemical to growing tissues, accumulate the glyphosate in the leaves first exposed to the chemical, and then shed just those leaves, leaving the rest of the plant free to grow. Other plant species produce a large amount of the enzyme that glyphosate targets. They do that by making extra copies of the gene that codes for the glyphosate-binding enzyme. With more enzyme than glyphosate, the plant can swamp out the herbicide's effect. Still other species compartmentalize glyphosate into the vacuole, the equivalent of the plant cell's large hallway closet, keeping it away from the enzyme pathways. That 44 (and counting) weed species have developed resistance to glyphosate means that farmers need to find another way to control weeds. Glyphosate shouldn't affect animals and insects, since we don't make the enzyme that glyphosate targets, although recent studies show that glyphosate can impact honeybees because of its effect on the bees' microbiome. There is a split opinion about health and environmental consequences of glyphosate, but many agree that it's less problematic than other herbicides.[11]

Pesticides have enabled both a scale and a style of farming that clearly produces more food per acre, but with definite impacts on the environment. The 3-year crop rotations of Charlemagne's time, along with intensive 2-year rotations incorporating cattle and fodder that returned nutrients to the soil by growing legumes or grazing cattle and sheep, seems both innovative and archaic. After European settlers destroyed the lives and lifestyle of the Native Americans, they had more land to farm than farmers. They could turn over prairie sod, farm until the soils were exhausted, and head west to plow up more virgin acres. Large-scale mechanized farming, focused on single crop production, worked because we usually prioritized technology over land stewardship. As we developed pesticides and fertilizers to increase yields, we left the wisdom of rotations and promoting beneficial insects to control pests primarily to a small and specialized group of farmers, either organic farmers or those using both conventional and organic techniques but focusing on sustainability. And in our haste to develop GE plants, we ignored ecological common sense, developing varieties with single-gene resistance traits that have a built-in shelf life given the dynamic nature of biological life.

An obvious fix to those insect pests that have evolved a way around whatever resistance the plant may muster is to stack resistance traits. So GE plants contain more than one Bt protein, or have both Bt genes and herbicide-resistant genes added, slowing down the time it takes for the variety to become obsolete. Insect resistance to Bt has been reported, but the spread of those resistance genes is slowed in a couple of ways. There are more than 200 different Bt proteins, some with specificity for just a subset of insect species. And farmers are encouraged to intersperse plants lacking Bt genes in islands within the Bt-engineered crops. These islands create refuges for insects so that they don't become resistant to Bt, slowing the spread of Bt resistance as insects from the islands mate with the others. The percentage of farmers growing Bt-modified crops who actually create optimally sized and located refuges isn't clear.

With tools to insert genes into our crops, we have created variants that confer resistance to chemicals or pests. But most of the traits that could increase crop yields enough to meet food demands in 2050 aren't single-gene traits. The perfect plant to take us into GR2.0 would potentially tolerate increasing temperatures, grow well even under drought conditions, and, like our peas and lentils, form a symbiosis with bacteria that can fix nitrogen from the atmosphere. Or perhaps the GR2.0 plant would be a perennial plant that lives for 3 to 5 years, instead of an annual that needs to be reseeded every year. It would have long roots that could access water at depth; help build the soil in the way our prairie grasses helped develop the deep, rich topsoil of the tallgrass prairies; and at the same time produce as many grains as an annual plant. None of those dream traits can be generated by the addition of a single gene or even just a few genes.

Drought stress, for example, is different from herbicide or insect resistance in that even the definition of drought varies depending on location. A drought in Minnesota could occur under conditions with more precipitation than fields in northern Montana experience all year. And in any given location, a shortage of moisture in the spring affects plants differently than a shortage of moisture in midsummer. Plants adapt to the water conditions of their environment by managing a beautifully coordinated stream of water, connecting the moisture in the soil with the water vapor in the atmosphere. Water moves from the soil into the root, because the roots are drier than the soil. And plants, with their easy access to sunlight to make carbohydrates, secrete a thin layer of sugary gel that coats the roots at the growing tips, trapping water from the surrounding soil. Once the water is in the roots, it gets pulled up to the leaves through the narrow tubes of xylem. That pull on the column of water comes from water evaporating through tiny openings in the leaves. Those openings, called stomates, are how plants get carbon dioxide (CO_2) to build sugars during photosynthesis. If the soil is dry and plants become water stressed, stomates close to stop the evaporation. It's efficient from a water retention perspective, but it also cuts off CO_2 necessary for plants to grow. Hence a dilemma for the plant: keep your stomates open to take in CO_2 for growth but lose water, or conserve water and stop growing. For us, it's not a dilemma. We want large plants for more food, so we irrigate our crops. But for plants that must thrive under whatever water supply the environment provides, they manage water shortages in all kinds of creative ways.

But wheat is different. Most wheat in the developed world isn't irrigated. We grow wheat in areas that receive too little rainfall for most other crops. Before our ancestors ever thought about grinding its seeds, wheat had flourished in dry climates by avoiding drought altogether. Its seeds germinated with the first fall rains, sinking roots into moist soil to grow the first leaves, and then pausing growth when temperatures dropped. After waiting out a cool, wet winter, the plants started growing again when the temperatures warmed in the spring. With a jump-start on other species that wait to germinate until the spring, wheat could take advantage of the early season water and nutrients. It flowered and produced seeds in the early summer, before the temperatures skyrocketed. It adjusted its life cycle to grow when water was available, flowering and producing seeds before there was any conflict over water.

Some plants, instead of just avoiding drought stress by limiting their growth to the seasons with water, have modified how they photosynthesize to be more efficient with CO_2 and consequently conserve water. Wheat, together with most of the plants that cover the earth, uses the energy captured from photons of light to drive a pathway that starts by attaching CO_2 onto a two-carbon molecule. Hence the name C3 photosynthesis—the first step generates a three-carbon molecule. Since sunlight drives photosynthesis, plants open their stomates for CO_2 when the sun is shining, which is also when evaporation is at its highest. Some plants native to warm grasslands, including corn, sorghum, and sugarcane, have evolved a more efficient approach to photosynthesis. These plants, called C4 plants, still open their stomates during the day, but instead of using the CO_2 directly to make carbohydrates, they attach it to a three-carbon holder molecule (becoming a four-carbon molecule, hence the name C4). In this way, they accumulate CO_2 in their cells, and when the plants get water stressed and the leaf openings shut down, they can continue to access CO_2 by cleaving it off the four-carbon molecule. C4 plants can keep building carbohydrates even when their stomates are closed. The extra step of attaching the CO_2 to a holder molecule uses more energy, so C4 photosynthesis is only an advantage in hot and dry environments. In addition, there are structural differences in the leaf cells of C4 plants compared with C3 plants. They have physically separated the light-capturing step of photosynthesis from the CO_2-fixing step into two different cell types within the leaf. The shift from C3 to C4 photosynthesis involves gene regulation—shutting off or turning on the expression of genes depending on the cell type.

The evolution of C4 photosynthesis from C3 plants has happened more than 60 times in different lineages over the past 35 million years. It's still not clear how that change occurred, but by comparing the genome sequence between C4 corn and C3 rice genomes, scientists suggest that as many as 600 transposable element insertions have changed how genes were regulated, and affected C4 plants' capacity to concentrate CO_2. To engineer the development of a different photosynthetic pathway in wheat as a way to increase its capacity to grow more in warmer, drier climates would require an understanding of the regulation of gene expression responsible for differentiation of cell types and metabolic pathways. Clearly, the addition of single resistance genes, as in our herbicide-tolerant varieties, is in an entirely different realm from engineering complex traits associated with drought resistance.

GE wheat doesn't exist commercially; it's been developed by many private and public entities, but never released for commercial production. There are many reasons for that, but one of the strongest is the resistance from markets in Europe and Japan—and US farmers aren't willing to forgo the option of selling their wheat to foreign markets. Russia, Canada, and the United States were the top three wheat exporters in 2018, but the US portion of the international trade has been declining because of increased growth from other regions, particularly eastern Europe.[12] Thus, the potential loss of foreign markets that prefer to avoid GE wheat contributes to a farmer's decision about growing GE crops.

Additionally, while wheat is grown on more acres of land than other crops, it generates only 20% of the farm income of corn; it's not clear that it's economically viable to sell GE wheat seeds. Corn and soybeans, for example, grow in areas with higher precipitation, so they produce more per acre than wheat, and the crops are worth 4 to 5 times more per acre. With that, farmers can afford to pay more for seeds. There was also concern from farmers that growing an herbicide-resistant wheat could make it more difficult to manage their fields. Farmers in dryland (nonirrigated) wheat regions often leave fields fallow one year to store soil moisture and ensure a higher wheat yield the following year. The potential of having volunteer wheat come up in the fields that couldn't be controlled with herbicide during the fallow year was enough to dissuade some farmers from supporting GE wheat.

Biotechnology has increased our capacity to obtain higher yields of some crops. But that has come at an expense: higher seed costs and herbicide-resistant weeds that complicate weed management for farmers. The traits that would be the most useful for increasing yield—drought tolerance, salt tolerance, and heat tolerance—aren't easily manipulated with the addition or subtraction of single genes. And the genomes of our food crops are complicated and messy enough that engineering complex traits seems difficult at best. Given all that, our capacity to produce enough food over the next generations can't depend solely on biotechnological developments, as advances made to date don't suggest that we're on the verge of GR2.0, a second period of yield increases like that which accompanied the changes in agricultural methods to include fertilizers and irrigation. That shifts the conversation to the need to think more holistically about agroecosystems so that we manage our pests in ways that don't impact beneficial species; address efficiency of food distribution to limit waste; support the diversity of farming systems to include large industrial and smaller farms; and put our technological expertise to work developing agricultural methods with less impact on agricultural and neighboring ecosystems.

CHAPTER NINE

A Love-Hate Relationship

Wheat has come a long way since its early days on the scattered oak hillslopes. It grows now on all but the polar continents, and is a staple food for a large portion of the world's population. But success attracts haters, not much different for plants than for people. Oftentimes for people, it's their very best traits that they're hated for; it's not so different for wheat. The most vociferous complaint against wheat is related to that attribute which brought this plant such attention, from early civilizations down to the present: its protein structure. Wheat was so much more popular than rye or oats among the elites of Egypt and of Rome, and the moneyed gentry of Europe throughout the Middle Ages and into the industrial age, because of the softness of white bread. With a hard outer crust and a light, porous interior, loaves of this bread could be torn or sliced into pieces that soaked up the juices on the plates of those who could afford meat dinners. For the peasants, the hardy grains of rye and oats, and if the harvest was good, sometimes mixed with a little wheat, comprised the bulk of their calories. Rye made a heavy bread, not at all like wheat bread. But wheat was harder to grow in cool and wet temperate climates; rye and oats more dependably produced enough grain to last the farmers until the following year's harvest.

What makes wheat bread light and easy to slice is the gluten proteins that hold the crumbs together, forming a three-dimensional net that surrounds the starch molecules. But wheat doesn't have gluten so that we can make good bread. Its seeds are designed to assure the development of the next season's plant. At the base of the wheat seed is the embryo, with tissues that are the primordial roots and leaves of the future plant seedling. This future plantlet is the part of the kernel that we call wheat germ. Since it was built to be the next generation, it's the part of the wheat seed with the highest protein content: a quarter to a third of its weight is protein, and between 5 and 20% fats. The embryo is small relative to the rest of the seed, taking up only about 2% of the grain's weight. Most of the rest of the seed is the endosperm, primarily starch grains with some storage proteins and fats that will sustain the developing embryo and the newly germinated plant. When the seed takes up water to get the process of germination rolling, the cells around the embryo activate enzymes that break down the endosperm starch into sugars. The storage proteins, including gluten, are a source of nitrogen for the plantlet, important for building the molecular machinery of photosynthesis. Those storage proteins are the ones that brought this plant so much love and now so much hate.

Gluten is a nitrogen source for the germinating embryos not only for wheat but also for barley, rye, and oats. Gluten in wheat is not a single protein, but maybe a hundred or more similar proteins. Why we love the gluten family is that some of those proteins, the gliadins, give unbaked bread dough its capacity to hold together while changing shape, to be kneaded and punched down in the process of bread making, for example. The other half of those proteins, the glutenins, provide elasticity, which helps the dough stretch when yeast releases CO_2 in the process of leavening, and again in the oven as the heat expands the air in the dough's pores, creating a bread with a light texture. Why some hate the gluten family is that all these proteins have sections that are rich in two amino acids, proline and glutamine, which are resistant to breakdown. Normally, the digestive enzymes produced in our stomachs, the walls of our intestines, and our pancreas break down proteins in our diet into single amino acids or short chains of a couple of amino acids. The sections of the gluten proteins that are rich in proline or glutamine remain as relatively long chains of amino acids, not easily absorbed and used by our bodies.[1]

In part, gluten deserves its bad reputation. For up to 1% of the population, it destroys the inner intestinal walls, causing pain and limiting the capacity to absorb nutrients. For most of us, the indigestible amino acid chains, in combination with the cellulose of the cell walls, pass through our intestines as primarily undigested fiber. But for people with celiac disease, gluten affects the mucus-lined walls of the intestine, causing the mucus to thin. It also damages the tiny hairs that line the intestines and increase the surface area for nutrient absorption. Celiac disease occurs worldwide, and more frequently in women than in men, and in children than in older people. It affects people with a specific genetic predisposition, but not everyone carrying those genes will necessarily develop celiac disease. Why some do and others don't is still under investigation. But for those who do, the almost singular treatment is to avoid wheat, rye, barley, and for some, oats.

There's another 1% of us who are allergic to wheat. Allergies are a reaction by our immune system to rid something from the body. The runny nose and watery eyes that can purge the body of a virus or bacteria get turned on in some people by a harmless trigger—dust mites or pollen grains, for instance. With food allergies, the immune system evokes a response in the digestive tract—cramping, vomiting, or diarrhea. Other reactions include a skin rash or trouble breathing. A wheat allergy is a response to one of the many wheat proteins, not necessarily just a gluten protein. One wheat allergy in particular was noted during Roman times: baker's asthma, which results from inhaling wheat dusts and flour, and can be triggered by fungal contaminants in the flour or by one of the wheat proteins.

In recent years, the availability of gluten-free food has skyrocketed, and predictions are that it will hold an increasing share of global food markets. That's good news for people with celiac disease and wheat allergies, but the gluten-free market is supported by far more than just those two groups. Nonceliac gluten sensitivity was first described in the 1970s. By the early 2000s, books describing the symptoms and causes of gluten sensitivity and cookbooks for gluten-free diets were regularly on the *New York Times* best seller list. Gluten has been linked in the popular press to weight gain, bloating, diarrhea, abdominal pain, fatigue, headache, mental confusion, anxiety, and depression. The media attention no doubt increases the number of people who self-report gluten sensitivity.

Nonceliac gluten sensitivity afflicts between 0.5 and 13% of the people in regions of the globe whose inhabitants get a major portion of their nutrition from wheat. The huge discrepancy in percentages comes from the fact that nonceliac gluten sensitivity is self-reported. Unlike celiac disease and wheat allergies, there's no blood or biological marker to confirm the diagnosis. Therefore, the gold standard for confirming that diagnosis is a double-blind placebo control test. After following a gluten-free diet for at least four weeks, the patient uses a food additive for a week, followed by a gluten-free week, and then a week using a second additive. The patient, who doesn't know which additive contains gluten, reports any symptoms after the introduction of each additive. In a review of studies reporting on double-blind placebo control tests for patients who report gluten sensitivities, fewer than 20% of the patients describe negative symptoms in response to the gluten addition to their diet. And almost twice that many people had negative responses to the gluten-free additive.[2]

The discrepancy between self-reported symptoms and response to the double-blind test may be because there's something wrong with the test. For example, if the amount of gluten in the food additive is too low, someone with gluten sensitivity may not report symptoms; that individual could be sensitive only at higher doses. Or perhaps a one-week exposure isn't enough to provoke a sensitivity reaction. The other possibility is that some people may be sensitive to something in wheat, but it's not gluten. If people report discomfort with the no-gluten additive but not with the gluten additive, it could be a different protein that they're reacting to, or it may not be a protein at all.

Another potential culprit for people who report gluten sensitivity is a collection of small sugars, a group christened the FODMAPs: *F* for *fermentable*; *O*, *D*, and *M* describe the length of the sugars (oligo, di- or monosaccharides, meaning sugars that are composed of several, 2, or 1 main sugar unit); *A* for *and*; and *P* for *polyols*, which are short, sweet molecules that technically are alcohols instead of sugars, but still add a sweet taste to food.[3] Because FODMAPs are such a varied group of molecules, they're found in a lot of different foods: wheat, milk, legumes (peas, beans, and lentils), certain fruits (apples, apricots, watermelon, berries, peaches, and others), and a long list of vegetables and nuts. For people with irritable bowel syndrome, an ailment characterized by digestive tract discomfort, some of the FODMAPs are slow to get absorbed by the intestine, and instead are slowly fermented by the microbes in that organ, a process which leads to an accumulation of gases. Additionally, some of the smaller sugars can increase the uptake of water by the intestines, causing bloating. Eliminating FODMAPs from the diet can bring some people relief from gut discomfort, but another consequence of eliminating that many foods is that the nutritional composition of their diet takes a big hit.

Some argue that the rise in gluten sensitivity is because our breeding programs have changed wheat so much so that it's actually harmful for us now. The logical conclusion would be to switch to the ancient grains, emmer and einkorn, or to landrace varieties of wheat that haven't been bred for high yield. But wild wheats can have a higher protein concentration than our modern varieties. The protein content of the seeds of wild einkorn and emmer varieties ranges from 16 to 28%. When our ancestors selected the larger seeds to replant, they favored plants that could add more starch to the seeds, thereby diluting protein content. The protein in modern bread wheat is about 12 to 14% for hard wheats, grown specifically for bread, and 7 to 11% for soft wheats, grown for flour that makes light pastries. And when comparing protein content in seeds of landraces to the more widely distributed high-yielding varieties of wheat, the results are equally ambiguous. The protein content of wheat not only varies between varieties, but for some varieties it's affected by fertilizer addition (more nitrogen fertilizer leads to higher protein content) and by the timing and the amount of rainfall. If it rains a lot as the grains are filling, the grains are bigger but their protein content lower, in a dilution effect. Alternatively, low precipitation during the grain filling leads to a higher protein content, with the glutenin portion of the protein profile being the most changeable. For those with celiac disease, it's the presence of specific gliadin proteins that matters, and those, too, are sometimes higher in the ancient grains—leaving at most tenuous support for the argument that breeding has made wheat a less valuable food source for us.[4]

It's very possible to increase gluten content through breeding. When wheat breeders cross two promising varieties, the first filter is how well the seeds grow in the field. Farmers want wheat varieties that produce a high yield under the growing conditions of that region. But for farmers to sell their wheat to a mill, the miller looks for grain that produces good flour. Heavier grains mill into more flour, so grain size makes a difference to a miller. Likewise, the miller must sell the flour to a baker, and bakers have specific requirements. Higher protein content means better bread; but if the baker is making cakes and cookies, flour with a lower protein content is called for. Any new variety of wheat, therefore, must meet the requirements of the farmer to grow well, the miller to grind well, and the baker to optimize baked goods. That translates into an extensive evaluation for any breeding program.

Once a variety of wheat stands out as growing well in the field, its grains go to the test lab to be weighed and analyzed; mixed under standard conditions to test the properties of the dough; and then baked or boiled into the desired end product—loaves, biscuits, or noodles. When a breeder releases a new variety of wheat, usually after at least 10 years of developing and testing, it has gone through extensive trials in the field and in the baking lab. In the process of selecting grains that make the best bread dough or have the ultimate noodle traits, the breeders can select for strains of wheat with higher gluten content. But selecting a strain of high-gluten wheat that doesn't grow well is a dead end for farmer use; and selecting a high-yielding wheat that makes poor-quality bread is a dead end for bakers. Hence the selection process acts simultaneously on multiple traits that will serve the farmer, the miller, and the end user.[5]

Besides the problems with its proteins, another complaint that's been lodged against wheat is that it's an annual crop plant. That's another two strikes against it, then, being both an annual and a crop plant. Annual plants complete their whole life cycle, from germinating to flowering and setting new seeds, then dying, within a single year. Most of our crop plants are annuals, because annuals must produce good seeds every year, or they'll disappear from an area. In contrast, a perennial, which lives multiple years, can wait out a bad weather year and sink reserves into the roots for use the following season. It would be pointless for an annual plant to invest a lot of resources into its roots, when really all that counts is the seed production. But the roots of perennial plants grow deep into the soil and bring carbon and other resources to the microbes and small animals that keep soils alive and healthy.

And that leads directly into the consequences of changing wildlands into croplands. The change in North America began when pioneers plowed up native plant communities, particularly the grasslands of the Midwest. Those grasslands fed large herds of bison and other grazers. The plants native to these rich ecosystems had roots that contributed to the deep, fertile soils of the plains. By converting these lands into agriculture, a mixed-plant community was lost to large swaths of monocultures, and the soil lost a dense network of roots, replaced by short-lived shallow roots growing less than 6 months each year. The efficiency of monoculture industrial farming for food production per acre of land is indisputable, as is the destruction of ecosystem services. Prairies, despite the fact that we call them grasslands, contain more flowering broad-leaved plants than grasses, and support rich insect communities. Before we relied almost solely on chemicals to keep insect pests under control, farmers had understood that their best (insect) friends were enemies of their enemies. Native plant communities provide habitat for predators of the insects that eat our crops, and they also provide flower resources for pollinators. Those same plant communities and their accompanying rich soils absorb rainfall and snowmelt, regulating water flow into streams. So as we better understand the implications of large-scale farming practices on ecological systems, plants like wheat and corn get vilified as purveyors of ecological disasters.

Farmers had replaced the rich native plant communities with short-lived plants that need to be seeded each year, and they worked the soil every spring with sticks, stone hoes, ards, and finally metal plows. Then they left their fields bare between the harvest and the following seeding time. The principles of ecological systems that keep natural communities diverse and functioning at a high level begin with capturing solar energy. With that, our story loops back full circle to the evolution of plants and their move from aquatic to terrestrial environments. Over hundreds of millions of years, the action of plants—releasing organic acids to weather the rocks and dropping dead leaves and roots to add organic matter—slowly built the layer of soil that supports plant communities and all the terrestrial animals that depend on plants. By plowing up those plant communities, bare soils are exposed to wind and rain, disrupting the alchemic dance that connects sunlight to our dinner tables. This realization has inspired scientists, who for almost a century now have worked on converting some of our most popular crops to perennials.

The first time I heard Wes Jackson speak, about 10 years ago, I was floored that someone had the hubris to think they could start with an annual plant and make it a perennial. After more exposure to agricultural practices and ways of thinking, that initiative seems a little less incredible to me. I see it now as a deep reverence for the ingenuity of ecological systems combined with the workings of an adaptive mammalian brain that tries to solve problems. Changing a short-lived plant into a perennial plant isn't a trivial task. But that hasn't slowed down Jackson, an iconic scientist who has ruffled more than a few feathers during his long career, first at Sacramento State University and then as president of the Land Institute, a position from which he recently retired to devote more time to his writing. Jackson founded the Land Institute in 1976 as a nonprofit organization devoted to applying ecological knowledge to agriculture. His logic was pretty straightforward. Agriculture based on cultivated annual plants contributes to the destruction of the soil. Monocultures aren't ecologically viable communities. If instead our grains and cooking oils and legumes came from perennials, plants that could regrow after harvesting, we could save fossil fuels by not having to run tractors across the fields every year to prepare the soil and plant the seeds. After harvest, the plants would remain alive, still rooted in the ground, and ready to sprout back the following spring, much as our lawns green up after a winter of neglect. Weeds aren't such a problem if perennial plants are in the field, taking up the space and resources. And perennial plant systems are much better than annual ones at storing carbon in the soils; with that, we could potentially sequester CO_2 in soils that weren't being cultivated anew each year.

But Jackson's vision doesn't stop there. The Land Institute has been working on developing not only perennial wheat but perennial oil seeds and perennial legumes. For the farm field to simulate the ecological systems of the prairies, the institute also advocates growing polycultures featuring mixtures of different perennial plants. A polyculture with plants that mature at different times and different heights presents a challenge for harvesting, but one that can be addressed through the engineering and design of our harvesting equipment. While that's not a simple challenge, the greater challenge is introducing a longer life span into our crops. As with drought tolerance, a plant's living 3 or 4 years instead of one is not the result of a shift in a single gene or a couple of genes. The differences in longevity arise from the triggers to flowering and seed production, how the plant allocates the sugars it builds from CO_2, the structure and longevity of its roots, and when its cells are programmed to die. The challenge lies in navigating the tradeoff an organism faces for how to allocate its resources. Like our deciding whether to pay the rent or make the car payment, plants make decisions—but in their case life-and-death ones—about how to allocate their carbon. And the tradeoff between an annual and a perennial is deciding between funneling resources into seeds and funneling resources into more leaves and roots—sexual versus asexual reproduction. Perennials produce fewer seeds because they must invest in structures that will keep them alive over the next winter, ready to grow back leaf tissue when the spring weather returns. That also means they grow larger from one year to the next, and their seed production the first year is likely smaller than that of the second year.

There are a couple of ways to produce a perennial grain crop. We could start with a wild perennial wheatgrass and select for desirable food traits. Our ancestors chose annual species because they produce bigger seeds more reliably, but if we picked a perennial grass with reasonably sized seeds and just kept planting the biggest seeds, in time we could have a fairly good food plant. The Rodale Institute began that endeavor in 1983 with Jackson's encouragement, working with intermediate wheatgrass, *Thinopyrum intermedium*, a perennial grass species native to Europe and Asia. It was brought to North America in the 1930s as a forage plant, because it grows well and its relatively large grains have a high protein content. The program transferred to the Land Institute in 2003, and the work has expanded to include collaborators at the University of Minnesota and the University of Manitoba. Today, the flour from intermediate wheatgrass is commercially available as a specialty grain under the name Kernza. In 2017, General Mills announced that its subsidiary, Cascadian Farms, would buy this new grain from farmers—an important step in supporting the grain's commercialization.

As the beginning of a possible success story, Kernza is a work in progress. Our species began selecting annual wheat seeds for desirable traits 10,000 years ago; we've been doing the same for perennial wheatgrass since the early 1980s. It's a challenging process. For example, intermediate wheatgrass seeds of wild plants are about one-eighth the size of domestic wheat grains, although with selective breeding, the grains have doubled in size. The long-term goal is to breed a variety of wheatgrass bearing a grain up to half the size of domestic wheat. Because of their small size, each perennial wheatgrass grain has a higher proportion of seed coat compared to endosperm, meaning proportionately more fiber versus white flour after milling. The grains of intermediate wheatgrass also have a higher protein content (closer to 20%) and lower starch content (50 versus 65%) than grains of bread wheat. Moreover, that protein content has different proteins from those of bread wheat. The storage proteins that intermediate wheatgrass uses to bolster the growth of its embryos after germination include gliadins—the portion of the gluten that's problematic for people with celiac disease—but only small quantities of glutenins, the proteins that bind dough together. That translates to flour that is good for making flatbreads such as muffins, or, when mixed with conventional wheat flour, can be added to bread dough.

It's still very early in the breeding process for perennial wheatgrasses. Norman Borlaug accelerated the breeding of rust-resistant wheats in Mexico by growing a fall-seeded crop in the lowlands and an early summer–seeded crop in the highlands. But there's no option to speed up the breeding process with perennials. Because the second-year growth and grain yield may differ from the first year's, the breeder needs at least a couple of years in the field to know which plants to select. The development of a new variety of annual domestic wheat takes at least 10 to 15 years of breeding to introduce a trait from another variety. But what breeders are doing with intermediate wheatgrass is more involved—they're domesticating a wild perennial. That requires selection of a lot of different traits: seeds that stay on the stem at maturity instead of breaking off; larger seeds, which will also increase the starch content and lower the protein content; better winter survival; increased number of stems in the second year to increase the number of flowers; and on and on. The responsiveness of intermediate wheatgrass to selection over a relatively short period looks promising. And the investment of a major food company helps build a market so that farmers have a financial incentive to plant a grain that for the foreseeable future will be lower yielding than domestic wheat. It's an encouraging start.

The second approach to generating perennial grains is to

cross domestic wheat with wild perennial wheatgrasses. The first perennial wheat resulting from a cross between annual wheat and a wild perennial relative was bred in Montana and named MT-2. It's the product of a cross between durum wheat (28 chromosomes) and intermediate wheatgrass (42 chromosomes), the same species that breeders have been working to domesticate. MT-2 was released in 1987, a process requiring that the breeder present data on how the variety grows at locations across the state, and provide information that's relevant for farmers, millers, and bakers. In this case, the purpose of the release was to make seeds available for planting as a forage and as material for breeders. MT-2 could be used by any breeders interested in developing perennial wheat or using it in crosses with commercial wheat, not only for perennial genes but as a source of beneficial genes for resistance to many diseases, viruses, and fungi plaguing the annual wheats.

The current assortment of hybridized wheats is a motley group, but they are all the result of a cross between one of the domestic wheats and a wild perennial wheatgrass. The domestic wheats vary in chromosome numbers (einkorn has 14 chromosomes, durum and emmer have 28, and bread wheat has 42), as do the perennial wheatgrass species (the three most common species have either 14, 42, or 70 chromosomes).[6] The logic of the breeding program is that the domestic wheat has food characteristics that are desirable, whereas the perennial wheatgrass has the genes for a longer life history, likely scattered across all the seven base chromosomes that make up its genome. But the wheatgrass genome also includes genes that reduce the quality of the hybrid as a food plant: namely the grain size, the capacity of the grain to remain on the stem, the protein composition, and so forth. When breeders introduce a single trait from one variety of wheat to another, for example the dwarf wheat characteristics that Borlaug introduced to rust-resistant wheat varieties in Mexico, they backcross the progeny to the original parent multiple times, saving seeds only from individuals that have the desirable traits of the original parent and the one trait they were searching for in the second variety. Breeding a perennial wheat is a challenge because that trait is coded for by a lot of different genes, and the wild wheatgrass also has traits that aren't desirable for food production. It's like walking a tightrope—falling off the right side of the rope leaves you without enough of the wheatgrass traits that confer perennial status to bread wheat. Falling off the left side of the rope leaves you with a perennial wheat that carries too many traits that reduce its value as a food crop.

Perennial wheat is not on the verge of replacing fields of bread wheat. But that doesn't devalue the breeding endeavor. Even if perennial crops aren't about to dominate food production, they provide a useful tool for farmers who want to increase the health of their soils. The earliest crop rotations developed in Sumerian times, if not before, were based on the observation that growing a nitrogen-fixing plant in the field could increase the growth of the wheat crop the next year. As our methods for studying soils become increasingly sophisticated, we know that in addition to the nutrients that we harvest in our crops, we are losing soil organic matter, the carbon from decomposing roots and leaves that is a source of food for soil communities and also a source of nutrients for the soil. Perennial crops, grown in rotation with annual cash crops, can help soils recover from intensive farming practices. When coupled with a legume, either in polyculture or by growing the wheat first and the legume afterward, the perennial crop can help rebuild soils by increasing the carbon that moves from the plant into the soil, add to the nitrogen supply of the soil, and protect the soil surface from the erosion that comes after clearing and tilling the fields.

In the early days of agriculture, grinding grass seeds for food and growing those same grasses—enough to provide food for the village, the first states, the citizens of a nation at war—were monumental tasks, with a record of both great successes and devastating failures. Contrary to Malthus's prediction in the eighteenth century that our population would grow exponentially (he wasn't altogether wrong on that point) while our food supply can increase only arithmetically, we have cracked the nut of food production on many different fronts. Our food supply has increased to the point that food crises as the result of crop failures have been remedied to a large extent, thanks mainly to global markets and our capacity to transport food where shortages exist. Recent famines have occurred because of intentional disruption of the food supply, not because our farmers can't grow enough food.

Yet while we have built agricultural systems that are impressively productive, we have disrupted ecological processes that we wrote off as inessential or irrelevant. The problems with ignoring basic ecological tenets are underscored by environmental crises and human health impacts. We have polluted waters and soils around the globe with an excess of the chemicals we add to the fields where we grow our food. Dead zones in the Gulf of Mexico are caused by nutrients that run off our farm fields into the Mississippi River. When those nutrients reach the Gulf waters, they cause an initial burst of growth that uses up the oxygen, choking out all life. Beneficial insects are disappearing, with shortages of native pollinators for our orchards, and now diebacks in honeybee colonies that we transport to agricultural lands to provide pollination services. And the incidence of cancer and other health problems has increased for people who work regularly with pesticides, because of exposure to chemicals in levels that their bodies can't process.

With what we have learned since we first started planting seeds and harvesting food—about how water and nutrients cycle and energy flows through ecosystems, how plant communities interact with the pathogens and herbivores that depend on them, how species evolve and adapt to changing environments—we could be doing a much better job of growing food within an ecological framework. That would mean incorporating at a larger scale the practices some farmers already use: rotating crops, diversifying fields, keeping soils healthy, and scattering nonagricultural land amid our fields as habitat for animals and insects that are part of a healthy ecological system. But how do you inspire changes in practices, particularly if those changes might mean lower short-term financial gains, and run counter to recommendations of global industries, which have prospered by marketing products that make farming easier and more productive? Farmers make decisions with an eye to markets and regulatory requirements. For a crop like wheat, those markets are often regional or international, rarely local. As long as ecology is separate from food production, and long-term ecological costs aren't included in the calculations of short-term economic gains, we can continue to grow crops with blinders on.

EPILOGUE

An Eternal Harvest

The first plants that colonized land 400 million years ago changed the earth's surface by breaking apart the rocks into a substrate that slowly developed into soil. They changed the trajectory of life by adding oxygen to the atmosphere to create a chemistry that favored the evolution of animals, including us. Of all the different species of plants—almost 400,000—there are an impressively small handful of species that we use as food.

Wheat possesses characteristics that since the early days of our relationship have ensured its place as one of the most important plants for our species. Its seeds are nutritious and large compared with the other grasses. Their capacity for storage was a preadaptation for us; we used those seeds as a source of currency, a commodity to be transported long distances and traded for other foods or luxuries, and a staple food for people around the world.

The cuisines that developed have generated a delectable variety of ways to cook or bake with wheat. Buttery layers of croissants, hard-crusted artisanal loaves of whole-grain bread, thick crackers, steamed buns, light chapati loaves, pasta of all shapes, Asian noodles, sweet baklava, flaky piecrusts, tabouleh, bulgur pilaf, burbara and other wheat berry porridges, pita bread, dumplings, matzos, pancakes, cookies, muffins, wontons, pizzas, fritters, biscuits—the list of ways that we use wheat grains could fill pages. People who suffer from celiac disease and so must be aware of all the places wheat or gluten is used as a food additive could add to that list—things like soy sauce and gravies that we don't automatically associate with wheat. We have adapted this grain in so many ways to satisfy our stomachs.

If the way we use wheat differs around the globe, so does the way we grow it. In the United States, our farms are larger. The average size for an American farm is over 400 acres; in Europe the average size is 40 acres. Yet globally, half the food calories produced are from farms of 12 acres or smaller. Large-acreage American farms mean that fewer of us put effort into growing food, and fewer of our children have a relationship with the plants we grow for food.

One of the benefits of food awareness is the potential for us to be more mindful of our food. At a time when many of our most pressing health issues are related to diet—heart disease, diabetes, certain cancers, and stroke—such mindfulness is welcome. But that's only a first step toward addressing malnutrition in our children: around the world, 265 million children under the age of 5 suffer from the effect of poor nutrition, including obesity. Ultimately, we're left with the complexities of managing our own food, in the context of global food patterns. The global significance of wheat also means that local control of this crop is more tenuous. There is a budding movement for making bread with locally grown and milled flour, but at least for now that movement is primarily focused on higher-end bakeries supported by foodies. The local food movement is in part a response to concerns about a global food system that is out of our control, but has had limited success with addressing food quality for people living at or below the poverty level in our country.

This biography started in the Fertile Crescent and ended in pantries and cupboards worldwide. It's the story of one plant and how we've used that plant, although some would argue that the story is how that plant used us to take over the world. Its story isn't entirely unique; we could tell a similar tale about rice and corn. The details of each grain are different, but together they comprise most of the caloric intake for humans.

The story of wheat is a remarkable account of the ingenuity of our own species to manage our food, of the adaptability of grasses to new soils and new climates, and of the intricacy of the ecological networks on which our relationship with our food plants depends. Our relationship with wheat isn't simple; managing food production isn't simple. In the United States, we use a combination of market pressures and government regulations and subsidies to make sure that we produce adequate quantities of our main foods. But markets rarely integrate long-term costs of ecological damage, and the list of the negative environmental impacts of intensive agriculture gets longer each year. That's where government policy to encourage more ecologically sustainable practices could have an impact. But those kinds of changes are dependent on political whims and winds, and influenced by consumers, by lobbyists for corporate agribusiness, and by foreign markets.

And the future for wheat? The question remains as to whether the adaptation and acclimation of wheat species and our other food crops can keep up with the changing pace of environmental conditions. Wheat grows in semiarid regions, and is adapted to dry growing seasons. But changes in the timing of precipitation and warmer temperatures can affect the growth of wheat and the success of the pathogens and herbivores that prey on it. We have moved wheat across continents, bred wheat by crossing varieties with different traits. With our increasing expertise in adding genes to and editing genes from genomes, we build the technological tools to keep modifying our food species. Except that the sequencing of the wheat genome underscores the circuitous pathways of evolution. If we were, as an intellectual exercise, to design a set of chromosomes and genes for a plant that would become a major crop species, it's less than unlikely that we would come up with the existing configuration of the wheat genome. Future research will help us understand the function of all that DNA that doesn't code for genes, and the evolutionary processes that have enabled us to move this plant around the globe. The wheat genome is a beautiful example of complexity that gives these plants the plasticity to respond to environmental change.

While it's easy to promise technological fixes for growing food for future generations, we are still subject to the laws of nature, of drought and water shortages, of soil erosion and the long soil-building processes, of sunlight and biochemical pathways that build sugars and organic matter. We live in an increasingly fast-paced intellectual world on a planet that operates on geologic timescales with ecological processes fueled by sunshine and evolution. We operate as though we are outside the laws of nature, but we're not. We're human. That's a lovely, beautiful, challenging, difficult thing. Wheat is wheat. That's an equally lovely, beautiful, challenging, and difficult thing. Together, we share a common goal: an eternal harvest.

The roots, the stems, the leaves, and the seeds of wheat and every other plant species have a story for us. Plants are just plants. Wheat gets unfairly blamed for many of our own shortcomings. But suppose for a minute that the whispering of the wheat stalks in the field had a story for us, one we could hear and learn from. That story might be about sun, about rain, about soil. It would be a fairly simple story—wheat plants don't have a brain, after all. But maybe that simple story could guide some of our policy that affects agricultural practices. Manage your soils as though they were a valuable resource. Grow your food with humanitarian principles in mind. Use your resources wisely. As a species, we have a stunning capacity for creativity and problem solving. Imagine if we focused all that capacity on optimizing agricultural production in the most environmentally sustainable way. Imagine what we could do.

Acknowledgments

This book came to be with help and support from many people. My agent, Jessica Papin, believed in my capacity to tell a good story, and has artfully guided me through the writing process, from the book proposal to the final draft. Christie Henry, then at University of Chicago Press, and Scott Gast, Michaela Luckey, Sandra Hazel, and Mary Corrado are part of the editorial team that saw my manuscript through the production stages.

Margaret Bryan Davis coupled her deep interest in ecological systems with integrity and creativity. I am a better scientist for having been part of her lab. While training in science is largely about learning to write technically and concisely, the late Eila Perlmutter introduced me to writing techniques to advance the story, whether that story is technical or lyrical or both.

The science of growing and breeding plants is complicated, and many people shared their time and knowledge with me. Many thanks to Wes Jackson, Lee DeHahn, Stan Cox, and Shu-wen Wang at the Land Institute; Scott McVey, Scott Bean, and Jeff Wilson at the USDA Agricultural Research Service; Dirk Maier and Jesse Poland at Kansas State University; Vicki Morrone at Michigan State University; and Robbin Moran at the New York Botanical Gardens. Luther Talbert at Montana State University answered my random questions and, along with Steve Fifield, gave me helpful comments on an early draft. Nate Powell-Palm, Randy Hinebauch, Bob Quinn, and Sam Schmidt shared their view of wheat from a farming and milling perspective.

Members of my writing group, especially Jan Davis, Kiki Rydell, Yoshi Colclough, Jia Hu, Ben Lei, Christine Lux, and Jan Zauha, provided feedback and encouragement when I was just starting to write this book.

I received a fellowship from the Arthur P. Sloan Foundation, and support from Montana State University for a sabbatical leave and a Faculty Excellence Grant.

Finally, my spouse, Rémy Jurie-Joly, was a sounding board for many, many ideas as I worked to translate technical information into an accessible story. His questions held me to a high standard of clarity.

Notes

CHAPTER ONE

1 Scientists have described a Great Oxidation Event (GOE), which occurred around 2 billion years ago. The GOE either arose contiguously with the first photosynthesizers or was delayed by the absorption of oxygen by volcanic gases. See the article by Lyons, Reinhard, and Planavsky, "The Rise of Oxygen in Earth's Early Ocean and Atmosphere."

2 It's possible to live your whole life without thinking about how the individual cells in your body know which part of the DNA they should turn on or shut off. But once you do start thinking about it, it's possible to obsess over it. That kind of obsession drives research programs.

3 The difficulty in knowing what caused glaciation cycles 444 million years ago is that our sampling is so sparse. We collect core samples that have some information relative to that time period, but without enough evidence preserved to piece together the whole story. Poroda et al., "High Potential for Weathering and Climate Effects of Non-Vascular Vegetation in the Late Ordovician," and Lenten et al., "First Plants Cooled the Ordivician," argue for the impact of weathering on Late Ordivician glaciation.

CHAPTER TWO

1 The recent finding in the Philippines of bones of what appears to be a new species of hominin, *Homo luzonensis*, exemplifies the discoveries related to the evolution of our species. The bones are

from the hand and foot, along with one leg bone and teeth. Those skeleton fragments have characteristics in common with both *Homo* and *Australopithecus*, as reported in Détroit et al., "A New Species of *Homo* from the Late Pleistocene of the Philippines."

2 The other shift that's not etched into the fossil record is physiological changes in our capacity to digest food, brought about by digestive enzymes.

3 The information about Abu Hureyra is summarized in *Village on the Euphrates*, a totally captivating volume by A. M. T. Moore, Gordon Hillman, and A. J. Legge, summarizing 25 years of analysis done on samples that had been collected before the dam was completed.

4 It's not clear when almonds were domesticated, but the Romans were eating almonds that they had raised which were not bitter, like their wild counterparts. The toxicity comes from a compound called amygdalin, common in other plants in the rose family. The pits of peaches, nectarines, apricots, and cherries contain this compound, which breaks down to release cyanide. Though we don't eat the pits of those fruits, we do eat that part of the seed in almonds. For more information, see Thodberg et al., "Elucidation of the Amygdalin Pathway Reveals the Metabolic Basis of Bitter and Sweet Almonds (*Prunus dulcis*)."

5 The saddle querns for grinding the seeds into flour had been used across the Levant and northern Africa for 10,000 years before the first mills were developed by the Mesopotamians, and perfected by the Romans in the last centuries BC.

6 Rye was the first domesticated grass at Abu Hureyra; it has the advantage over wheat of being much easier to prepare. Its lighter husk can be removed fairly easily with a mortar and pestle, and because it's light, the husk is more easily separated from the grain. Grinding rye into flour is faster, and it softens into a porridge more quickly than wheat. So, given its ease of preparation, you might wonder why this book is about wheat and not rye. The next evidence we find for domesticated rye comes from a site in southwestern Turkey 4,000 years later, and a rye granary from a settlement in north central Turkey dates to 4,000 years after that.

The rye grain that we grow for whiskey and pumpernickel bread and livestock feed is different from that which was found in the Levant. It originated in central Europe, where it was domesticated sometime between the sixth and the first centuries BC.

CHAPTER THREE

1 Managing river flows for irrigation is thought to have started around 5500 BC, and the irrigation systems of the Sumerians along the Euphrates River and the Egyptians along the Nile have been well studied. The villages throughout the Levant were often sited near springs or some other water source. Evidence of small-scale irrigation projects would be difficult to find in archaeological records, with the exception of a site in southern Jordan with buildings dated to 9,500 years ago and evidence of a cistern and, later, small dams (see Mithen, "The Domestication of Water").

2 Genetic variation in humans is as important as it is in any species; the story of King Charles II of Spain underscores that point by detailing the consequences of avoiding it. Charles was the last of the Hapsburg Dynasty, a family that wanted to keep the advantages of their royal lineage in the family. Charles's father, for example, married his niece, and was not the first in his lineage to make such a match. Other relatives married first cousins or first cousins once removed. Poor Charles, who was born in the mid-seventeenth century, was unable to walk until he was four or to speak before he reached the age of eight, not to mention unable to make decisions or sire children in either of his two marriages. Charles likely suffered from two rare, recessive disorders, recessive meaning that in order for the trait to be expressed, he had to have two copies of the gene coding for that trait, one from each parent. Mating between close relatives increases the likelihood that both parents will carry a copy of a rare gene; hence their children will inherit the required two copies of the gene coding for the disease. In this case, the Hapsburg bloodline could have benefited from an influx of new genes.

3 I have oversimplified this hybridization event, but here are more

details. Wild einkorn wheat (*Triticum monococcum*) split from its close relative *T. urartu* about a million years ago. It was *T. urartu* that crossed with a now extinct species of goatgrass, related to the modern-day *Aegilops speltoides*, that formed emmer. Scientists believe that there were two different hybridization events between those two species, one which formed emmer wheat (*Triticum turgidum*) and the other which formed *T. timopheevii*, both with 28 chromosomes. *T. timopheevii* is found only in the Transcaucasus region. This is summarized in Matsuoka, "Evolution of Polyploid Triticum Wheats under Cultivation."

4 Bread, porridge, and ale mark the development of a staple food for the diet of people of the Mediterranean region. On continental Europe, other than that which was grown for the Roman Empire, wheat was a minor crop. Farmers living far enough away from the Danube or the Rhone, thus outside the influence of Roman appetites, grew barley, rye, and oats instead of wheat. Wheat cultivation was minimal throughout much of Europe until the nineteenth century.

CHAPTER FOUR

1 Theophrastus, *Historia Plantarum* 8.8.1.

2 Legumes aren't the only plants that form symbioses with nitrogen-fixing bacteria. Some species of trees and flowering plants do the same, albeit with different bacterial symbionts. There are also free-living nitrogen fixers that dwell directly in soils, usually soils containing large amounts of organic matter. These bacteria acquire energy for splitting the triple bond of gaseous nitrogen by breaking chemical bonds in the organic matter.

3 In an experiment to see how long grains could be stored belowground, Ouafaa Kadim saw only a slight decline in the quality of wheat grains after 12 months of storage. See van Gign, Whittaker, and Anderson, *Exploring and Explaining Diversity in Agricultural Technology*.

CHAPTER FIVE

1 Islamic urban areas prospered beginning around 800, with 1.5 million people living in cities in North Africa and the Middle East, supported by a rich agricultural system. The golden age of Islam lasted from around the mid-seventh century up until the thirteenth century. But the key to the Islamic Empire's farming system was also a weak point. In warm and dry habitats, irrigation water gets taken up by the plant, and most of what is left in the soil evaporates, leaving a residue of salt that is dissolved in any natural water source. In a wet climate, rainfall washes the salt out of the soil. But in hot and arid climates, the salt can accumulate in the soil, until eventually the fields are nonproductive for anything but salt-tolerant plants.

2 For clarification of the Holy Roman Empire, Voltaire described it as neither Holy, nor Roman, nor an empire. Historians debate much about this period, including the timing of the beginning of the feudal system, or whether the system should even be called feudal. See Sassen, *Territory, Authority, Rights*, for a discussion of the feudal system as the precursor to the first nation-states.

3 Serfdom ended in Russia in 1861. The ideology supporting both slavery and its abolition is thoughtfully laid out in David Brion Davis's book, *The Problem of Slavery in Western Culture*.

4 This information comes from the work *Geoponika: Farm Work; A Modern Translation of the Roman and Byzantine Farming Handbook*, translated by Andrew Dalby and including essays on numerous farming topics.

5 Varro's writings can be found in *On Agriculture*, translated by W. D. Hooper and H. B. Ash, which includes works on agriculture by the Roman statesman Marcus Porcius Cato.

6 The yield ratios of wheat and other crops are derived from inventory stocks of manorial estates, the accuracy of which has been disputed. An underreporting of wheat production could result from tenants claiming lower harvests to reduce the taxable amount they needed to turn over to the manor. The yield of wheat crops was important, because grains weren't just a side dish for

medieval Europeans. They accounted for up to 60% of the diet, used in bread of course, but also in porridges and stews and definitely in ales. Rye was much more important than wheat for the peasants, because rye can tolerate poor soils and cold climates and is less susceptible to disease. A thick rye bread and porridge formed the basis of many peasants' diets. But wheat was a necessity for the nobility, as their diet included meats and fishes and spicy sauces, best when soaked up with a slice of light bread.

7 Sometimes referred to as the Big Death, the plague was introduced from Crimea and spread clockwise around Europe. A tiny rod-shaped cell, the bacteria *Yersinia pestis* spends most of its time in small rodents—squirrels or mice or rats—but it was wheat and our warring tendencies that helped spread the plague throughout Europe. Populations of the bacteria were centered in East Asia, far from European farm fields. One explanation for how the bacteria arrived in Europe was that a band of infected Mongol warriors attacked Italian merchants in a trading town in the Crimea in 1346. Shortly thereafter, both warriors and merchants got sick. Winter set in, and no one left port until the following spring, when the Italians returned home, carrying the bacteria. In 1347, the plague arrived in Constantinople, and crossed the Mediterranean to Crete and Sicily and Sardinia. The following year, people were dying from the plague throughout France, Italy, eastern Europe, and southern England. From there, the bacteria traveled in ships carrying grain through the North and Baltic Seas, along the trade routes to Oslo and northern Poland in 1349; even those ships that were empty but with enough scattered pockets of grain in their nooks and crannies to attract rodents carried the bacteria. In that way, the plague spread between the British Isles and eastern Europe. Inland, the bacteria moved through Germany and eastern Europe, and over the next few years across Russia. DNA capture from the teeth and bones of plague victims' corpses have been used to confirm *Yersinia pestis* as the cause of the plague. See Bos et al., "A Draft Genome of *Yersinia pestis* from Victims of the Black Death," and Haensch et al., "Distinct Clones of *Yersinia pestis* Caused the Black Death."

8 Pre-Columbian farming in the Americas was based on practices that included no livestock, as the only large mammals that had been domesticated were llamas and alpacas, pack animals better designed to carry loads than pull a plow. The Inca empire along the Andean *cordillera* (mountain range) supported itself with terraced farming on the hillslopes, fed by irrigation canals and fertilized with waste from the villages, or along the coast, fertilized with dung from seabirds. The archaeological evidence for farming practices in the humid tropical regions of South America is scant, but in some places it appears that raised-bed farming was employed. Soil was applied on top of wetland layers, keeping crops above the water table and providing an interface between the wetland and the soil for the delivery of nutrients—a practice unknown in our current agricultural systems.

9 Although there are various versions of the story of the Fifes, perhaps the most reliable comes from an essay in a collection by A. H. Reginald Buller, *Essays on Wheat.*

10 While some people attribute the Mennonites' success to emigrating with a superior variety of wheat, in *The Profit of Wheat* Courtney Fullilove makes a strong case that, accustomed as they were to agriculture on the Russian steppes, the Mennonites had in effect been managing their land for the challenges of prairie life, including fires and grasshopper plagues, thereby making Turkey a successful wheat variety.

CHAPTER SIX

1 Another word for "mineral" is *inorganic*, meaning that no carbon-based organic matter is included.

2 According to the author, Joseph LeConte, all that was needed to produce these salts was decaying organic matter, some soil, moisture and oxygen, and shelter from rain to prevent leaching of the salts. This manual was published in South Carolina, and its purpose was likely not to increase farm productivity but to stockpile ammunition during the Civil War.

3 Domestication, some would argue, is easily the greatest change

we have brought about on our crop plants; the rest of our breeding efforts have resulted in relatively small changes.

4 Darwin, *On the Origin of Species by Means of Natural Selection.*

5 See Wood et al., "Defining the Role of Common Variation in the Genomic and Biological Architecture of Adult Human Height," for a study of human height; and White and Rabago-Smith, "Genotype-Phenotype Associations and Human Eye Color," for a review of the genetic control of eye color.

CHAPTER SEVEN

1 The German government differentially distributed food stores to the military versus civilians, and the amount of food stored was not easy for people outside the government to ascertain. People also accessed food from the black market. Even so, tolerating food shortages during the war was one thing, but continued shortages after the Armistice were largely demoralizing for many. See Offer, *The First World War.*

2 The role of grain elevators and traders in the second half of the nineteenth century is described by William Cronon in *Nature's Metropolis*. Grain elevators functioned like banks, with wheat or corn or rye as the form of money, and receipts for grain delivered that could be bought and sold, with speculation on future prices. Groups of traders could buy up all the stocks of wheat, and force those who had invested in futures, and were bound to deliver that wheat at a future date, to purchase wheat from those who had "cornered" the market at whatever exorbitant price they demanded.

3 President Woodrow Wilson appointed Herbert Hoover to lead the US Food Administration. It was Hoover who pushed for establishing a minimum price for wheat along with promoting the Food Will Win the War conservation program.

4 This was not the first large-scale appropriation of reservation lands. In the late nineteenth century under the Dawes Act, reservation lands were divided by allotting 160- or 320-acre parcels of land to each member of a given tribe. The legislation was intended to encourage Native American self-sufficiency, similar to the ideolo-

gy underlying homesteading allotments to pioneers of the western United States. After tribal members received their allotment, on many reservations the rest of their land was opened for white homesteading, sometimes resulting in a patchwork of landownership. Besides losing lands to corporate farming initiatives, Indian tribes contributed to the war effort by enlisting in the armed services. Those who remained at home were equally involved: programs to encourage Indians to plow lands and plant seeds were supported during the war years and increased farm output.

5 Campbell reportedly paid around $340,000 for the original $2-million investment.

6 In wealthy countries, the number of people employed in farming is now less than 5%, while in less developed countries, it is closer to 60% (Roser, "Employment in Agriculture").

7 Quotation is from Campbell, *Campbell's 1905 Soil Culture Manual*, 8.

8 There were as many as 30,000 new land claims each year from the mid-1910s until the early 1920s.

9 Germany's chancellor from 1930 to 1932 was Heinrich Brüning, whose interest in paying off the reparations took priority over easing the effects of the Depression, exacerbating his nation's economic hardships.

10 In *The Taste of War*, Lizzie Collingham argues that the pressing need to provide food contributed to the haste with which people were murdered in mass graves and later in the concentration camps. Even providing them with small amounts of food diverted calories that could have been going to the military.

11 Congressional funding of the Soil Conservation Service came just after a presentation to Congress by Hugh Bennett, then the director of the recently formed Soil Erosion Service. Upon hearing that a dust storm originating in New Mexico was on its way to Washington, DC, Bennett prolonged his presentation until the topsoil darkened the skies of the nation's capital—a most dramatic argument for the need for better soil management.

12 See Helfer, "Rust Fungi and Global Change," for a recent review on rusts and climate change, and Wright, "Chinch Bug Management," for an overview of chinch bugs.

13 "Green Revolution" was used in a speech given in 1968 by William Gaud, director of the US Agency for International Development, and may be the first public use of that term.

CHAPTER EIGHT

1 Only 5% of the developed world's wheat production was on irrigated lands, as reported in 1989, in contrast to the situation in developing countries, where wheat is more commonly grown on irrigated lands, 90% of that in Africa (Sayre, "Management of Irrigated Wheat").

2 In keeping with the role of foundations in food aid, the Bill and Melinda Gates Foundation has been a key driver of the conversation about GR 2.0. See Pingali, "The Green Revolution."

3 See International Wheat Genome Sequencing Consortium, "Shifting the Limits in Wheat Research and Breeding Using a Fully Annotated Reference Genome."

4 In "Chromosome Organization and Genic Expression," Barbara McClintock describes the identification of transposable elements; Nina Federof, who researched the same system with modern techniques, provides an overview of McClintock's work in "McClintock's Challenge in the 21st Century." The discovery of a similar process caused by viruses infecting bacteria is described by Austin Taylor in "Bacteriophage-Induced Mutation in *Escherichia coli*." For a biography of McClintock, see Evelyn Fox Keller, *A Feeling for the Organism*.

5 With the utmost respect for air traffic controllers, I used this analogy only because the quality of their work is so high that we fly in airplanes feeling safe, which would never happen if there wasn't a well-organized traffic coordination system. Genetic systems must be regulated with similar precision.

6 See Sinzelle, Izsvák, and Ivics, "Molecular Domestication of Transposable Elements"; Alzohairy et al., "Transposable Elements Domesticated and Neofunctionalized by Eukaryotic Genomes"; and Liang et al., "Kicking against the PRCs," for examples of transposable elements being modified into functional genes. Some scien-

tists think this is still rather speculative, but others (see Federoff, "Transposable Elements, Epigenetics, and Genome Evolution," and Shapiro, "Living Organisms Author Their Read-Write Genomes in Evolution") see more potential for interactive genomes, in part as a result of the activity of transposable elements.

7 There are a series of dwarf genes that have different effects on not just height but other plant traits, including the number of seed heads produced. The 24 currently identified height-regulating genes (Rht) are on different chromosomes. For example, two Rht genes are carried by semidwarf Norin 10, one on the fourth chromosome from the first goatgrass that hybridized with a relative of einkorn wheat, and the other on the fourth chromosome of the second goatgrass that hybridized with emmer wheat. Rht-24 is the most common dwarf gene in Chinese varieties, and is found on the sixth chromosome of the einkorn relative (Tian et al., "Preliminary Exploration of the Source, Spread, and Distribution of Rht24 Reducing Height in Bread Wheat").

8 See USDA Economic Research Service, "Recent Trends in GE Adoption: Adoption of Genetically Engineered Crops in the U.S.," for data from the United States; and ISAAA, "Biotech Crop Annual Update, 2018," for global data.

9 The other two herbicides for which there are resistant GE crops are glufosinate and dicamba, with more in development. Authors dispute whether there is a real increase in productivity with GE crops, but two meta-analyses providing support for that idea are Klümper and Qaim, "A Meta-analysis of the Impacts of Genetically Modified Crops," and Pellegrino et al., "Impact of Genetically Engineered Maize on Agronomic, Environmental and Toxicological Traits."

10 Glyphosate is assumed to be relatively benign for mammals and insects, since we don't produce those same amino acids and thus aren't susceptible to the pattern of glyphosate disruption.

11 The list of herbicide-resistant weeds can be found in Heap, "International Survey of Herbicide Resistant Weeds." Eric Patterson and others, "Glyphosate Resistance and EPSPS Gene Duplication," describe multiple gene copies conferring resistance to glypho-

sate, while Alejandro Perez-Jones and others, "Investigating the Mechanisms of Glyphosate Resistance in *Lolium multiflorum*," and Nick Harre and others, "Differential Antioxidant Enzyme Activity in Rapid-Response Glyphosate-Resistant *Ambrosia trifida*," discuss other mechanisms of resistance. Glyphosate effects on bee colonies are reported in Motta, Raymann, and Moran, "Glyphosate Perturbs the Gut Microbiota of Honey Bees."

12 Export data was extracted from Workman, "Wheat Exports by Country."

CHAPTER NINE

1 Rye, barley, and oats have their own version of gluten proteins that are equally indigestible, just not in the same quantity as wheat. If a wheat grain averages 12% protein, the grains of rye, barley, and oats are 7 to 8% protein.

2 See Molina-Infante and Carroccio, "Suspected Nonceliac Gluten Sensitivity Confirmed in Few Patients after Gluten Challenge in Double-Blind, Placebo-Controlled Trials," and Francaville et al., "Randomized Double-Blind Placebo-Controlled Crossover Trial for the Diagnosis of Non-celiac Gluten Sensitivity in Children," for information on double-blind placebo-controlled tests for gluten sensitivity.

3 The sugar you might add to your coffee or tea is sucrose, which is made up of two sugar units: a glucose and a fructose. Sucrose, therefore, is a disaccharide, whereas glucose and fructose are monosaccharides.

4 In "Presence of Celiac Disease Epitopes in Modern and Old Hexaploid Wheat Varieties, " Hetty Van der Broeck and others argue that breeding has contributed to an increase in gliadin proteins; in "New Insights into Wheat Toxicity," Miguel Ribeiro and others conclude that when growing all the wheat varieties in the same soils with fertilizer, those differences disappear. Perhaps a more significant source of increased gluten in our diet is its use as a food additive. Because the whole-grain portion of whole-wheat flour reduces the capacity of the dough to make a light-textured

bread, additional gluten, either added to the flour by the miller or added to the dough by the baker, strengthens the protein matrix enough to support the whole-wheat materials. In "Do Ancient Types of Wheat Have Health Benefits Compared with Modern Bread Wheat?," Peter Shewry reviews studies that report on both the potential health benefits and the adverse health effects of ancient grains and modern varieties, and concludes that more research is needed to reconcile studies with contrasting effects.

5 Glutenin proteins affect both the elasticity of the dough and the volume of a loaf of bread, and are divided into high-molecular-weight (HMW) and low-molecular-weight (LMW) proteins. The genes for glutenin proteins are on the first chromosome of each of the three sets of chromosomes that bread wheat carries. The genes from the second goatgrass parent, what wheat breeders call the D-genome (A coming from the einkorn relative, B from the first goatgrass parent, and D from the second goatgrass parent), have two superior alleles (gene forms), designated as 5+10, which produce flour that makes a high-volume loaf. Breeders of bread wheat will start with a variety that carries genes for 5+10 HMW glutenin and cross it with a variety that has other desirable traits. That's how breeders control gluten content without using genetic engineering techniques. See F. MacRitchhie, "Seventy Years of Research into Breadmaking Quality," and Devinder Mohan and Raj Gupta, "Gluten Characteristics Imparting Bread Quality in Wheats Differing for High Molecular Weight Glutenin Subunits at Glu D1 Locus."

6 Researchers from Washington State University have been working with crosses of bread wheat with tall wheatgrass. The most successful perennial lines from those crosses have somewhere between 48 and 62 chromosomes. At the Land Institute, MT-2 survives, but has rarely produced grain in Kansas fields. So scientists there made their own crosses between durum wheat and intermediate wheatgrass. The perennial wheat hybrids that Chinese scientists work with have between 50 and 56 chromosomes, with the successful lines carrying at least 6 to 8 chromosomes from the perennial wheatgrass.

Bibliography

CHAPTER ONE

Blankenship, Robert. "Early Evolution of Photosynthesis." *Plant Physiology* 154, no. 2 (2010): 434–38.

Chanderbali, André S., Brent A. Berger, Dianella G. Howarth, Pamela S. Soltis, and Douglas E. Soltis. "Evolving Ideas on the Origin and Evolution of Flowers: New Perspectives in the Genomic Era." *Genetics* 202, no. 4 (2016): 1255–65.

Chanderbali, André S., Mi-Jeong Yoo, Laura M. Zahn, Samuel F. Brockington, P. Kerr Wall, Matthew A. Gitzendanner, Victor A. Albert, et al. "Conservation and Canalization of Gene Expression during Angiosperm Diversification Accompany the Origin and Evolution of the Flower." *Proceedings of the National Academy of Sciences of the United States of America* 107, no. 52 (2010): 22570–75.

Ciaffi, Mario, Anna Paolacci, Oronzo Tanzarella, and Enrico Porceddu. 2011. "Molecular Aspects of Flower Development in Grasses." *Sexual Plant Reproduction* 24, no. 4 (2011): 247–82.

De Clerck, Olivier, Kenny A. Bogaert, and Frederik Leliaert. "Diversity and Evolution of Algae: Primary Endosymbiosis." *Advances in Botanical Research* 64 (2012): 55–86.

Fangel, Jonatan U., Peter Ulvskov, J. P. Knox, Maria D. Mikkelsen, Jesper Harholt, Zoë A. Popper, and William G.T. Willats. "Cell Wall Evolution and Diversity." *Frontiers in Plant Science* 3 (2012): 1–8.

Galtier, J., and F. M. Hueber. "How Early Ferns Became Trees." *Proceedings of the Royal Society B: Biological Sciences* 268, no. 1479 (2001): 1955–57.

Gensel, Patricia G., and Christopher M. Berry. "Early Lycophyte Evolution." *American Fern Journal* 91, no. 3 (2001): 74–98.

Ghienne, Jean-François, André Desrochers, Thijs R. A. Vandenbroucke, Aicha Achab, Esther Asselin, Marie-Pierre Dabard, Claude Farley, Alfredo Loi, Florentin Paris, Steven Wickson, and Jan Veizer. "A Cenozoic-Style Scenario for the End-Ordovician Glaciation." *Nature Communications* 5 (2014). https://doi.org/10.1038/ncomms5485.

Harholt, Jesper, Øjvind Moestrup, and Peter Ulvskov. "Why Plants Were Terrestrial from the Beginning." *Trends in Plant Science* 21, no. 2 (2015): 96–101.

Harrison, C. Jill, and Jennifer L. Morris. "The Origin and Early Evolution of Vascular Plant Shoots and Leaves." *Philosophical Transactions of the Royal Society of London. Series B, Biological Sciences* 373, no. 1739 (2018). https://doi.org/10.1098/rstb.2016.0496.

Hedges, S. Blair, Qiqing Tao, Mark Walker, and Sudhir Kumar. "Accurate Timetrees Require Accurate Calibrations." *Proceedings of the National Academy of Sciences* 115 (2018): E9510–11. https://doi.org/10.1073/pnas.

Heo, Jung-Ok, Pawel Roszak, Kaori M. Furuta, and Ykä Helariutta. "Phloem Development: Current Knowledge and Future Perspectives." *American Journal of Botany* 101, no. 9 (2014): 1393–402.

Herron, Matthew D., Jeremiah D. Hackett, Frank O. Aylward, and Richard E. Michod. "Triassic Origin and Early Radiation of Multicellular Volvocine Algae." *Proceedings of the National Academy of Sciences of the United States of America* 106, no. 9 (2009): 3254–58.

Jones, Victor A. S., and Liam Dolan. "The Evolution of Root Hairs and Rhizoids." *Annals of Botany* 110, no. 2 (2012): 205–12.

Kenrick, Paul, and Peter R. Crane. "The Origin and Early Evolution of Plants on Land." *Nature* 389, no. 6676 (1997): 33–39.

Kenrick, Paul, and Christine Strullu-Derrien. "The Origin and Early Evolution of Roots." *Plant Physiology* 166, no. 2 (2014): 570–80.

Leliaert, Frederik, Heroen Verbruggen, and Frederick W. Zechman. "Into the Deep: New Discoveries at the Base of Green Plant Phylogeny." *BioEssays* 33, no. 9 (2011): 683–92.

Lenton, Timothy M., Michael Crouch, Martin Johnson, Nuno Pires, and Liam Dolan. "First Plants Cooled the Ordovician." *Nature Geoscience* 5 (2012): 86–89.

Ligrone, Roberto, Jeffrey G. Duckett, and Karen. S. Renzaglia. "Conducting Tissues and Phyletic Relationships of Bryophytes." *Royal Society* 355, no. 1398 (2000): 795–813.

Ligrone, Roberto, Jeffrey G. Duckett, and Karen S. Renzaglia. "Major Transitions in the Evolution of Early Land Plants: A Bryological Perspective." *Annals of Botany* 109, no. 5 (2012): 851–71.

Linkies, Ada, Kai Graeber, Charles Knight, and Gerhard Leubner-Metzger. "The Evolution of Seeds." *New Phytologist* 186, no. 4 (2010): 817–31.

Lucas, William J., Andrew Groover, Raffael Lichtenberger, Kaori Furuta, Shri-Ram Yadav, et al. "The Plant Vascular System: Evolution, Development and Functions." *Journal of Integrative Plant Biology* 55, no. 4 (2013): 294–388.

Lyons, Timothy W., Christopher T. Reinhard, and Noah J. Planavsky. "The Rise of Oxygen in Earth's Early Ocean and Atmosphere." *Nature* 506, no. 7488 (2014): 307–15.

Mathews, Sarah, and Elena M. Kramer. "The Evolution of Reproductive Structures in Seed Plants: A Re-examination Based on Insights from Developmental Genetics." *New Phytologist* 194, no. 4 (2012): 910–23.

Morris, Jennifer L., Mark N. Puttick, James W. Clark, Dianne Edwards, Paul Kenrick, et al. "Reply to Hedges et al.: Accurate Timetrees Do Indeed Require Accurate Calibrations." *Proceedings of the National Academy of Sciences* 115, no. 4 (2018): E9512–13.

Morris, Jennifer L., Mark N. Puttick, James W. Clark, Dianne Edwards, Paul Kenrick, et al. "The Timescale of Early Land Plant Evolution." *Proceedings of the National Academy of Sciences* 115, no. 10 (2018): E2274–83.

Niklas, Karl J. "The Cell Walls That Bind the Tree of Life." *BioScience* 54, no. 9 (2004): 831–41.

Niklas, Karl J. "The Evolutionary-Developmental Origins of Multicellularity." *American Journal of Botany* 101, no. 1 (2014): 6–25.

Panchy, Nicholas, Melissa Lehti-Shiu, and Shin-Han Shiu. "Evolution of Gene Duplication in Plants." *Plant Physiology* 171, no. 4 (2016): 2294–316.

Pigg, Kathleen B. "Isoetalean Lycopsid Evolution: From the Devonian to the Present." *American Fern Journal* 91, no. 3 (2001): 99–114.

Pittermann, J. "The Evolution of Water Transport in Plants: An Integrated Approach." *Geobiology* 8, no. 2 (2010): 112–39.

Popper, Zoë A., Gurvan Michel, Cécile Hervé, David S. Domozych, William G. T. Willats, Maria G. Tuohy, Bernard Kloareg, and Dagmar B. Stengel. "Evolution and Diversity of Plant Cell Walls: From Algae to Flowering Plants." *Annual Review of Plant Biology* 62 (2011): 567–90.

Porada, P., T. M. Lenton, A. Pohl, B. Weber, L. Mander, Y. Donnadieu, C. Beer, U. Pöschl, and A. Kleidon. "High Potential for Weathering and Climate Effects of Non-Vascular Vegetation in the Late Ordovician." *Nature Communications* 7 (2016):12113. https://doi.org/10.1038/ncomms12113.

Prochnik, Simon E., James Umen, Aurora M. Nedelcu, Armin Hallmann, Stephen M. Miller, et al. "Genomic Analysis of Organismal Complexity in the Multicellular Green Alga *Volvox carteri*." *Science* 329, no. 5988 (2010): 223–26.

Quirk, Joe, Jonathan R. Leake, David A. Johnson, Lyla L. Taylor, Loredana Saccone, and David J. Beerling. "Constraining the Role of Early Land Plants in Palaeozoic Weathering and Global Cooling." *Proceedings of the Royal Society B: Biological Sciences* 282 (2015). https://doi.org/10.1098/rspb.2015.1115.

Sarkar, Purbasha, Elena Bosneaga, and Manfred Auer. "Plant Cell Walls throughout Evolution: Towards a Molecular Understanding of Their Design Principles." *Journal of Experimental Botany* 60, no. 13 (2009): 3615–35.

Shaw, Jonathan, and Karen Renzaglia. "Phylogeny and Diversification of Bryophytes." *American Journal of Botany* 91, no. 10 (2004): 1557–81.

Sørensen, Iben, David Domozych, and William G. T. Willats. "How Have Plant Cell Walls Evolved?" *Plant Physiology* 153, no. 2 (2010): 366–72.

Sørensen, Iben, Filomena A. Pettolino, Antony Bacic, John Ralph, Fachuang Lu, Malcolm A. O'Neill, Zhangzhun Fei, Jocelyn K. C. Rose, David S. Domozych, and William G. T. Willats. "The Charophycean Green Algae Provide Insights into the Early Origins of Plant Cell Walls." *The Plant Journal* 68, no. 2 (2011): 201–11.

Stein, William E., Christopher M. Berry, Linda VanAller Hernick, and Frank Mannolini. "Surprisingly Complex Community Discovered in the Mid-Devonian Fossil Forest at Gilboa." *Nature* 483, no. 7387 (2012): 78–81.

Strömberg, Caroline A. E. "Evolution of Grasses and Grassland Ecosystems." *Annual Review of Earth and Planetary Sciences* 39 (2011): 517–44.

Umen, James G. "Green Algae and the Origins of Multicellularity in the Plant Kingdom." *Cold Spring Harbor Perspectives in Biology* 6, no. 11 (2014). https://doi.org/10.1101/cshperspect.a016170.

Vanholme, Ruben, Brecht Demedts, Kris Morreel, John Ralph, and Wout Boerjan. "Lignin Biosynthesis and Structure." *Plant Physiology* 153, no. 3 (2010): 895–905.

Vasco, Alejandra, Robbin C. Moran, and Barbara A. Ambrose. "The Evolution, Morphology, and Development of Fern Leaves." *Frontiers in Plant Science* 4 (2013). https://doi.org/10.3389/fpls.2013.00345.

Weng, Jing-Ke, and Clint Chapple. "The Origin and Evolution of Lignin Biosynthesis." *New Phytologist* 187, no. 2 (2010): 273–85.

CHAPTER TWO

Attwell, Laura, Kris Kovarovic, and Jeremy Kendal. "Fire in the Plio-Pleistocene: The Functions of Hominin Fire Use, and the Mechanistic, Developmental and Evolutionary Consequences." *Journal of Anthropological Sciences* 93 (2015). doi 10.4436/JASS.93006.

Banks, William E., Nicolas Antunes, Solange Rigaud, and Francesco D'Errico. "Ecological Constraints on the First Prehistoric Farmers in Europe." *Journal of Archaeological Science* 40, no. 6 (2013): 2746–53.

Belfer-Cohen, Anna, and A. Nigel Goring-Morris. "Becoming Farmers: The Inside Story." *Current Anthropology* 52, no. S4 (2011): S209–20.

Belfer-Cohen, Anna, and Erella Hovers. "The Ground Stone Assemblages of the Natufian and Neolithic Societies in the Levant—a

Brief Review." *Journal of the Israel Prehistoric Society* 35 (2005): 299–308.

Bellwood, Peter S. "The Beginnings of Agriculture in Southwest Asia." In *First Farmers: The Origins of Agricultural Societies*, 44–66. Malden, MA: Blackwell, 2005.

Blockley, S. P. E., and R. Pinhasi. "A Revised Chronology for the Adoption of Agriculture in the Southern Levant and the Role of Lateglacial Climatic Change." *Quaternary Science Reviews* 30, nos. 1–2 (2011): 98–108. https://doi.org/10.1016/j.quascirev.2010.09.021.

Bocquet-Appel, Jean-Pierre. "When the World's Population Took Off: The Springboard of the Neolithic Demographic Transition." *Science* 333, no. 6042 (2011): 560–61.

Carmody, R. N., G. S. Weintraub, and R. W. Wrangham. "Energetic Consequences of Thermal and Nonthermal Food Processing." *Proceedings of the National Academy of Sciences* 108, no. 48 (2011): 19199–203.

Charmet, Gilles. "Wheat Domestication: Lessons for the Future." *Comptes Rendus Biologies* 334, no. 3 (2011): 212–20.

Détroit, Florent, Armand Salvador Mijares, Julien Corny, Guillaume Daver, Clément Zanolli, Eusebio Dizon, Emil Robles, Rainer Grün, and Philip J. Piper. "A New Species of *Homo* from the Late Pleistocene of the Philippines." *Nature* 568, no. 7751 (2019): 181–86.

Eitam, David, Mordechai Kislev, Adiel Karty, and Ofer Bar-Yosef. "Experimental Barley Flour Production in 12,500-Year-Old Rock-Cut Mortars in Southwestern Asia." *PLoS One* 10, no. 7 (2015): e0133306. https://doi.org/10.1371/journal.pone.0133306.

Fernández, Eva, Alejandro Pérez-Pérez, Cristina Gamba, Eva Prats, Pedro Cuesta, Josep Anfruns, Miquel Molist, Eduardo Arroyo-Pardo, and Daniel Turbón. "Ancient DNA Analysis of 8000 B.C. Near Eastern Farmers Supports an Early Neolithic Pioneer Maritime Colonization of Mainland Europe through Cyprus and the Aegean Islands (Ancient DNA Analysis of Near Eastern Farmers)." *PLOS Genetics* 10, no. 6 (2014): e1004401. https://doi.org/10.1371/journal.pgen.1004401.

Fiorenza, Luca, Stefano Benazzi, Amanda G. Henry, Domingo C. Salazar-García, Ruth Blasco, Andrea Picin, Stephen Wroe, and Ottmar Kullmer. "To Meat or Not to Meat? New Perspectives on

Neanderthal Ecology." *American Journal of Physical Anthropology* 156 (2015): 43–71.

Fu, Qiaomei, Cosimo Posth, Mateja Hajdinjak, Martin Petr, Swapan Mallick, et al. "The Genetic History of Ice Age Europe." *Nature* 534, no. 7606 (2016): 200–205.

Fuller, Dorian Q., George Willcox, and Robin G. Allaby. "Early Agricultural Pathways: Moving outside the 'Core Area' Hypothesis in Southwest Asia." *Journal of Experimental Botany* 63, no. 2 (2012): 617–33.

Haak, Wolfgang, Peter Forster, Barbara Bramanti, Shuichi Matsumura, Guido Brandt, Marc Tänzer, Richard Villems, et al. "Ancient DNA from the First European Farmers in 7500-Year-Old Neolithic Sites." *Science* 310, no. 5750 (2005): 1016–18.

Hardy, Karen, Jennie Brand-Miller, Katherine D. Brown, Mark G. Thomas, and Les Copeland. "The Importance of Dietary Carbohydrate in Human Evolution." *Quarterly Review of Biology* 90, no. 3 (2015): 251–68.

Humphrey, Louise T., Isabelle De Groote, Jacob Morales, Nick Barton, Simon Collcutt, Christopher Bronk Ramsey, and Abdeljalil Bouzouggar. "Earliest Evidence for Caries and Exploitation of Starchy Plant Foods in Pleistocene Hunter-Gatherers from Morocco." *Proceedings of the National Academy of Sciences of the United States of America* 111, no. 3 (2014): 954–59.

Kılınç, Gülşah Merve, Ayça Omrak, Füsun Özer, Torsten Günther, Ali Metin Büyükkarakaya, et al. "The Demographic Development of the First Farmers in Anatolia." *Current Biology* 26, no. 19 (2016): 2659–66.

Larson, Greger, Dolores R. Piperno, Robin G. Allaby, Michael D. Purugganan, Leif Andersson, et al. "Current Perspectives and the Future of Domestication Studies." *Proceedings of the National Academy of Sciences of the United States of America* 111, no. 17 (2014): 6139–46.

Lazaridis, Iosif, Dani Nadel, Gary Rollefson, Deborah C. Merrett, Nadin Rohland, et al. "Genomic Insights into the Origin of Farming in the Ancient Near East." *Nature* 536, no. 7617 (2016): 419–24.

Li, Chunxiang, Diane L. Lister, Hongjie Li, Yue Xu, Yinqiu Cui, Mim

A. Bower, Martin K. Jones, and Hui Zhou. "Ancient DNA Analysis of Desiccated Wheat Grains Excavated from a Bronze Age Cemetery in Xinjiang." *Journal of Archaeological Science* 38, no. 1 (2011): 115–19.

Luca, F., G. H. Perry, and A. Di Rienzo. "Evolutionary Adaptations to Dietary Changes." *Annual Review of Nutrition* 30 (2010): 291–314.

Mariotti Lippi, Marta, Bruno Foggi, Biancamaria Aranguren, Annamaria Ronchitelli, and Anna Revedin. "Multistep Food Plant Processing at Grotta Paglicci (Southern Italy) around 32,600 cal B.P." *Proceedings of the National Academy of Sciences of the United States of America* 112, no. 39 (2015): 12075–80.

Molleson, Theya, Karen Jones, and Stephen Jones. "Dietary Change and the Effects of Food Preparation on Microwear Patterns in the Late Neolithic of Abu Hureyra, Northern Syria." *Journal of Human Evolution* 24, no. 6 (1993): 455–68.

Moore, A. M. T., Gordon C. Hillman, and A. J. Legge. *Village on the Euphrates: From Foraging to Farming at Abu Hureyra*. With contributions by J. Huxtable, M. Le Mière, T. I. Molleson, D. de Moulins, S. L. Olsen, D. I. Olszewski, V. Roitel, et al. London: Oxford University Press, 2000.

Pinhasi, Ron, Mark G. Thomas, Michael Hofreiter, Mathias Currat, and Joachim Burger. "The Genetic History of Europeans." *Trends in Genetics* 28, no. 10 (2012): 496–505.

Piperno, Dolores R., Ehud Weiss, Irene Holst, and Dani Nadel. "Processing of Wild Cereal Grains in the Upper Palaeolithic Revealed by Starch Grain Analysis." *Nature* 430, no. 7000 (2004): 670–73.

Prasad, Vandana, Caroline A. E. Strömberg, Habib Alimohammadian, and Ashok Sahni. "Dinosaur Coprolites and the Early Evolution of Grasses and Grazers." *Science* 310, no. 5751 (2005): 1177–80.

Price, T. Douglas. "Ancient Farming in Eastern North America." *Proceedings of the National Academy of Sciences of the United States of America* 106, no. 16 (2009): 6427–28.

Purugganan, Michael D., and Dorian Q. Fuller. "The Nature of Selection during Plant Domestication." *Nature* 457, no. 7231 (2009): 843–48.

Revedin, Anna, Biancamaria Aranguren, Roberto Becattini, Laura Longo, Emanuele Marconi, Marta Mariotti Lippi, Natalia Skakun, Andrey Sinitsyn, Elena Spiridonova, and Jirí Svoboda. "Thirty-Thousand-Year-Old Evidence of Plant Food Processing." *Proceedings of the National Academy of Sciences of the United States of America* 107, no. 44 (2010): 18815–19.

Rollefson, Gary O., Alan H. Simmons, and Zeidan Kafafi. "Neolithic Cultures at 'Ain Ghazal, Jordan." *Journal of Field Archaeology* 19, no. 4 (1992): 443–70.

Salamini, Francesco, Hakan Özkan, Andrea Brandolini, Ralf Schäfer-Pregl, and William Martin. "Genetics and Geography of Wild Cereal Domestication in the Near East." *Nature Reviews Genetics* 3, no. 6 (2002): 429–41.

Sanchez, Miguel Cortes, Francisco J. Jimenez Espejo, Maria D. Simon Vallejo, Juan F. Gibaja Bao, Antonio Faustino Carvalho, et al. "The Mesolithic–Neolithic Transition in Southern Iberia." *Quaternary Research* 77, no. 2 (2012): 221–34.

Sikora, Martin, Andaine Seguin-Orlando, Vitor C. Sousa, Anders Albrechtsen, Thorfinn Korneliussen, et al. "Ancient Genomes Show Social and Reproductive Behavior of Early Upper Paleolithic Foragers." *Science* 358, no. 6363 (2017): 659–62.

Speth, John D. "When Did Humans Learn to Boil?" *PaleoAnthropology* 2015:54–67.

Tattersall, Ian. "Human Evolution and Cognition." *Theory in Biosciences* 129, no. 2 (2010): 193–201.

Tattersall, Ian. "If I Had a Hammer." *Scientific American* 311, no. 3 (2014): 55–59.

Thodberg, Sara, Jorge Del Cueto, Rosa Mazzeo, Stefano Pavan, Concetta Lotti, Federico Dicenta, Elizabeth H. Jakobsen Neilson, et al. "Elucidation of the Amygdalin Pathway Reveals the Metabolic Basis of Bitter and Sweet Almonds (*Prunus dulcis*)." *Plant Physiology* 178, no. 3 (2018): 1096–111.

Tresset, Anne, and Jean-Denis Vigne. "Last Hunter-Gatherers and First Farmers of Europe." *Comptes Rendus Biologies* 334, no. 3 (2011): 182–89.

Weiss, Ehud, and Daniel Zohary. "The Neolithic Southwest Asian Founder Crops: Their Biology and Archaeology." *Current Anthropology* 52, no. 5 (2011): S237–54.

CHAPTER THREE

Alvarez, Gonzalo, Francisco C. Ceballos, and Celsa Quinteiro. "The Role of Inbreeding in the Extinction of a European Royal Dynasty (Inbreeding in a Royal Dynasty)." *PLoS One* 4, no. 4 (2009): e5174. https://doi.org/10.1371/journal.pone.0005174.

Bernstein, William J. *A Splendid Exchange: How Trade Shaped the World.* New York: Atlantic Monthly Press, 2008.

Betts, Alison, Peter Weiming Jia, and John Dodson. "The Origins of Wheat in China and Potential Pathways for Its Introduction: A Review." *Quaternary International* 348 (October 2014): 158–68.

Cook, L. M., B. S. Grant, I. J. Saccheri, and J. Mallet. "Selective Bird Predation on the Peppered Moth: The Last Experiment of Michael Majerus." *Biology Letters* 8, no. 4 (2012): 609–12.

Dvorak, Jan, Karin R. Deal, Ming-Cheng Luo, Frank M. You, Keith von Borstel, and Hamid Dehghani. "The Origin of Spelt and Free-Threshing Hexaploid Wheat." *Journal of Heredity* 103, no. 3 (2012): 426–41.

Kellogg, Elizabeth A. "The Evolutionary History of Ehrhartoideae, Oryzeae, and Oryza." *Rice* 2, no. 1 (2009): 1–14.

Kellogg, Elizabeth A. "Relationships of Cereal Crops and Other Grasses." *Proceedings of the National Academy of Sciences of the United States of America* 95, no. 5 (1998): 2005–10.

Kramer, Samuel N. *The Sumerians: Their History, Culture, and Character.* Chicago: University of Chicago Press, 1963.

Laghetti, Gaetano, Girolamo Fiorentino, Karl Hammer, and Domenico Pignone. "On the Trail of the Last Autochthonous Italian Einkorn (*Triticum monococcum* L.) and Emmer (*Triticum dicoccon* Schrank) Populations: A Mission Impossible?" *Genetic Resources and Crop Evolution* 56, no. 8 (2009): 1163–70.

Longin, C. Friedrich H., and Tobias Würschum. "Back to the Future—

Tapping into Ancient Grains for Food Diversity." *Trends in Plant Science* 21, no. 9 (2016): 731–37.

Madella, Marco, Juan García-Granero, Welmoed Out, Philippa Ryan, and Donatella Usai. "Microbotanical Evidence of Domestic Cereals in Africa 7000 Years Ago." *PLoS One* 9, no. 10 (2014). https://doi.org/10.1371/journal.pone.0110177.

Matsuoka, Yoshihiro. "Evolution of Polyploid Triticum Wheats under Cultivation: The Role of Domestication, Natural Hybridization and Allopolyploid Speciation in Their Diversification." *Plant and Cell Physiology* 52, no. 5 (2011): 750–64.

Mithen, Steven. 2010. "The Domestication of Water: Water Management in the Ancient World and Its Prehistoric Origins in the Jordan Valley." *Philosophical Transactions of the Royal Society A: Mathematical, Physical and Engineering Sciences* 368, no. 1931 (2010): 5249–74.

Peng, Junhua, Dongfa Sun, and Eviatar Nevo. "Wild Emmer Wheat, *Triticum dicoccoides*, Occupies a Pivotal Position in Wheat Domestication Process." *Australian Journal of Crop Science* 5, no. 9 (2011): 1127–43.

Samuel, Delwen. "Experimental Grinding and Ancient Egyptian Flour Production." In *Beyond the Horizon: Studies in Egyptian Art, Archaeology and History in Honour of Barry J. Kemp*, edited by S. Ikram and A. Dodson, 456–77. Cairo: Publications of the Supreme Council of Antiquities, 2009.

Simmons, Alan. "Mediterranean Island Voyages." *Science* 338, no. 6109 (2012): 895–97.

Spengler, Robert, Michael Frachetti, Paula Doumani, Lynne Rouse, Barbara Cerasetti, Elissa Bullion, and Alexei Mar'yashev. "Early Agriculture and Crop Transmission among Bronze Age Mobile Pastoralists of Central Eurasia." *Proceedings of the Royal Society B: Biological Sciences* 281, no. 1783 (2014). https://doi.org/10.1098/rspb.2013.3382.

Styring, Amy K., Michael Charles, Federica Fantone, Mette Marie Hald, Augusta McMahon, et al. "Isotope Evidence for Agricultural Extensification Reveals How the World's First Cities Were Fed."

Nature Plants 3, no. 6 (2017). https://doi.org/10.1038/nplants.2017.76

Unal, H. Güran. "Some Physical and Nutritional Properties of Hulled Wheat." *Tarim Bilimleri Dergisi* 15, no. 1 (2009): 58–64.

van't Hof, Arjen E., Pascal Campagne, Daniel J. Rigden, Carl J. Yung, Jessica Lingley, Michael A. Quail, Neil Hall, Alistair C. Darby, and Ilik J. Saccheri. "The Industrial Melanism Mutation in British Peppered Moths Is a Transposable Element." *Nature* 534, no. 7605 (2016): 102–5.

van't Hof, Arjen E., Nicola Edmonds, Martina Dalíková, Frantisek Marec, and Ilik J. Saccheri. "Industrial Melanism in British Peppered Moths Has a Singular and Recent Mutational Origin." *Science* 332, no. 6032 (2011): 958–60.

Zhou, Yong, Zhongxu Chen, Mengping Cheng, Jian Chen, Tingting Zhu, Rui Wang, Yaxi Liu, et al. "Uncovering the Dispersion History, Adaptive Evolution and Selection of Wheat in China." *Plant Biotechnology Journal* 16, no. 1 (2018): 280–91.

CHAPTER FOUR

Bogaard, Amy. "The Archaeology of Food Surplus." *World Archaeology* 49, no. 1 (2017): 1–7.

Diamond, Jared. *Guns, Germs and Steel: The Fates of Human Societies*. New York: W. W. Norton, 1997.

Gurevitch, J., S. M. Scheiner, and G. A. Fox. *The Ecology of Plants*. 2nd ed. Sunderland, MA: Sinauer Associates, 2006.

Halstead, Paul. *Two Oxen Ahead: Pre-mechanized Farming in the Mediterranean*. Sussex, UK: Wiley-Blackwell, 2014.

Isett, Christopher, and Stephen Miller. *The Social History of Agriculture*. Lanham, MD: Rowman and Littlefield, 2017.

Kadim, Ouafaa. "A Participatory Approach to Post-Harvest Loss Assessment: Underground and Outdoor Cereal Storage in Doukkala, Morocco." In *Exploring and Explaining Diversity in Agricultural Technology*, edited by Annelou van Gign, John Whittaker, and Patricia C. Anderson, 199–203. Oxford: Oxbow Books, 2014.

Kennett, Douglas J., and James P. Kennett. "Early State Formation in Southern Mesopotamia: Sea Levels, Shorelines, and Climate

Change." *Journal of Island and Coastal Archaeology* 1, no. 1 (2006): 67–99.

Kohler, Timothy A., Michael E. Smith, Amy Bogaard, Gary M. Feinman, Christian E. Peterson, et al. "Greater Post-Neolithic Wealth Disparities in Eurasia Than in North America and Mesoamerica." *Nature* 551, no. 7682 (2017): 619–22.

Mann, Michael. *The Sources of Social Power.* Vol. 1, *A History of Power from the Beginning to AD 1760*. Cambridge: Cambridge University Press, 2012.

Mazoyer, Marcel, and Laurence Roudart. *A History of World Agriculture: From the Neolithic Age to the Current Crisis.* New York: Monthly Review Press, 2006.

Roberts, J. M. *The New Penguin History of the World.* London: Penguin Books, 2007.

Scott, James C. *Against the Grain: A Deep History of the Earliest States.* New Haven, CT: Yale University Press, 2017.

Tauger, Mark B. *Agriculture in World History*. Edited by Peter N. Stearns. Themes in World History. London: Routledge, 2011.

Urem-Kotsou, Dushka. "Storage of Food in the Neolithic Communities of Northern Greece." *World Archaeology* 49, no. 1 (2017): 73–89.

van Gign, Annelou, John Whittaker, and Patricia C. Anderson. *Exploring and Explaining Diversity in Agricultural Technology*. Oxford: Oxbow Books, 2014.

Weiss, H., M.-A. Courty, W. Wetterstrom, F. Guichard, L. Senior, R. Meadow, and A. Curnow. "The Genesis and Collapse of Third Millennium North Mesopotamian Civilization." *Science* 261, no. 5124 (1993): 995–1004.

White, K. D. *Roman Farming*. Edited by H. H. Scullard. Aspects of Greek and Roman Life. Ithaca, NY: Cornell University Press, 1970.

CHAPTER FIVE

Andersen, Thomas Barnebeck, Peter Sandholt Jensen, and Christian Volmar Skovsgaard. "The Heavy Plow and the Agricultural Revolution in Medieval Europe." *Journal of Development Economics* 118 (January 2016): 133–49.

Benedictow, Ole J. "The Black Death." *History Today* 55, no. 3 (2005): 42–49.

Bonjean, Alain P., and William J. Angus. *The World Wheat Book: A History of Wheat Breeding*. Paris: Lavoisier, 2001.

Bos, Kirsten I., Verena J. Schuenemann, G. Brian Golding, Hernán A. Burbano, Nicholas Waglechner, Brian K. Coombes, Joseph B. McPhee, et al. "A Draft Genome of *Yersinia pestis* from Victims of the Black Death." *Nature* 478, no. 7370 (2011): 506–10.

Bosker, Maarten, Eltjo Buringh, and Jan Luiten van Zanden. "From Baghdad to London: Unraveling Urban Development in Europe, the Middle East, and North Africa, 800–1800." *Review of Economics and Statistics* 95, no. 4 (2013): 1418–37.

Brereton, John. "Briefe and True Relation of the Discoverie of the North Part of Virginia in 1602." Published in 1602. Madison: American Journeys Collection, Wisconsin Historical Society Digital Library and Archives.

Buller, A. H. Reginald. *Essays on Wheat, including the Discovery and Introduction of Marquis Wheat, the Early History of Wheat-Growing in Manitoba, Wheat in Western Canada, the Origin of Red Bobs and Kitchener, and the Wild Wheat of Palestine*. New York: Macmillan, 1919.

Cato, and Varro. *On Agriculture*. Translated by W. D. Hooper and H. B. Ash. Loeb Classical Library, no. 283. Cambridge, MA: Harvard University Press, 1934.

Crosby, Alfred W., Jr. *The Columbian Exchange: Biological and Cultural Consequences of 1492*. 30th anniversary ed. Westport, CT: Praeger, 2003.

Dalby, Andrew, trans. *Geoponika: Farm Work; A Modern Translation of the Roman and Byzantine Farming Handbook*. Totnes, Devon: Prospect Books, 2011.

Davis, David Brion. *The Problem of Slavery in Western Culture*. Ithaca, NY: Cornell University Press, 1966.

Epstein, Steven A. *An Economic and Social History of Later Medieval Europe, 1000–1500*. Cambridge: Cambridge University Press, 2009.

Frank, Robert Worth, Jr. "The Fourteenth-Century Agricultural Crisis." In *Agriculture in the Middle Ages: Technology, Practice, and Representation*, edited by Del Sweeney, 227–44. Philadelphia: University of Pennsylvania Press, 1995.

Fullilove, Courtney. *The Profit of the Earth: The Global Seeds of American Agriculture*. Chicago: University of Chicago Press, 2017.

Gráda, Cormac Ó, and Jean-Michel Chevet. "Famine and Market in ancien régime France." *The Journal of Economic History* 62, no. 3 (2002): 706–33.

Haensch, Stephanie, Raffaella Bianucci, Michel Signoli, Minoarisoa Rajerison, Michael Schultz, Sacha Kacki, Marco Vermunt, et al. "Distinct Clones of *Yersinia pestis* Caused the Black Death." *PLoS Pathogens* 6, no. 10 (2010): e1001134. https://doi.org/10.1371/journal.ppat.1001134.

Hoffman, Richard C. *An Environmental History of Medieval Europe*. Cambridge: Cambridge University Press, 2014.

Hourani, Albert. *A History of the Arab Peoples*. Cambridge, MA: Belknap Press of Harvard University Press, 1991.

Jordan, William Chester. *The Great Famine: Northern Europe in the Early Fourteenth Century*. Princeton, NJ: Princeton University Press, 1996.

Kingsbury, Noel. *Hybrid: The History and Science of Plant Breeding*. Chicago: University of Chicago Press, 2009.

Lenz, Kristina, and Nils Hybel. "The Black Death: Its Origin and Routes of Dissemination." *Scandinavian Journal of History* 41, no. 1 (2016): 54–70.

Mann, Charles C. *The Wizard and the Prophet: Two Remarkable Scientists and Their Dueling Visions to Shape Tomorrow's World*. New York: Alfred A. Knopf, 2018.

Mann, Michael. *The Sources of Social Power*. Vol. 1, *A History of Power from the Beginning to AD 1760*. Cambridge: Cambridge University Press, 2012.

Morris, Craig F. "Puroindolines: The Molecular Genetic Basis of Wheat Grain Hardness." *Plant Molecular Biology* 48 (2002): 633–47.

Murphy, Denis J. *People, Plants, and Genes*. Oxford: Oxford University Press, 2007.

Olmstead, Alan L., and Paul W. Rhode. *Creating Abundance: Biological Innovation and American Agricultural Development*. Cambridge: Cambridge University Press, 2008.

Phillips, William D. *Slavery from Roman Times to the Early Transatlantic Trade*. Minneapolis: University of Minnesota Press, 1985.

Reif, J. C., P. Zhang, S. Dreisigacker, M. L. Warburton, M. van Ginkel, D. Hoisington, M. Bohn, and A. E. Melchinger. "Wheat Genetic Diversity Trends during Domestication and Breeding." *Theoretical and Applied Genetics* 110, no. 5 (2005): 859–64.

Renard, D., J. Iriarte, J. J. Birk, S. Rostain, B. Glaser, and D. McKey. "Ecological Engineers Ahead of Their Time: The Functioning of Pre-Columbian Raised-Field Agriculture and Its Potential Contributions to Sustainability Today." *Ecological Engineering* 45 (2012): 30–44.

Revel, Jean-Francois. *Culture and Cuisine: A Journey through the History of Food*. Translated by Helen R. Lane. Garden City, NY: Doubleday, 1982.

Rosen, William. *The Third Horseman*. New York: Viking, 2014.

Rubin, Miri. *The Middle Ages: A Very Short Introduction*. Oxford: Oxford University Press, 2014.

Sassen, Saskia. *Territory, Authority, Rights: From Medieval to Global Assemblages*. Princeton, NJ: Princeton University Press, 2006.

Van Bavel, Bas. "Early Proto-industrialization in the Low Countries? The Importance and Nature of Market-Oriented Non-agricultural Activities on the Countryside in Flanders and Holland, c. 1250–1570." *Revue Belge de Philologie et d'Histoire* 81, no. 4 (2003): 1109–65.

Watson, Andrew M. "The Arab Agricultural Revolution and Its Diffusion, 700–1100." *Journal of Economic History* 34, no. 1 (1974): 8–35.

White, Lynn, Jr. *Medieval Technology and Social Change*. London: Oxford University Press, 1962.

Wickham, Chris. *Medieval Europe*. New Haven, CT: Yale University Press, 2016.

Zinn, Howard. *A People's History of the United States*. New York: Harper and Row, 1980.

CHAPTER SIX

Barnum, Dennis W. "Some History of Nitrates." *Journal of Chemical Education* 80, no. 12 (2003): 1393–96.

Bull, Alan T., Juan A. Asenjo, Michael Goodfellow, and Benito Gómez-Silva. "The Atacama Desert: Technical Resources and the Growing

Importance of Novel Microbial Diversity." *Annual Review of Microbiology* 70, no. 1 (2016): 215–34.

Carpenter, Daniel P. *The Forging of Bureaucratic Autonomy: Reputations, Networks, and Policy Innovation in Executive Agencies, 1862–1928*. Edited by Ira Katznelson, Martin Shefter, and Theda Skocpol. Princeton, NJ: Princeton University Press, 2001.

Christenhusz, Maarten J. M., Rafaël Govaerts, John C. David, Tony Hall, Katherine Borland, Penelope S. Roberts, Anne Tuomisto, Sven Buerki, Mark W. Chase, and Michael F. Fay. "Tiptoe through the Tulips—Cultural History, Molecular Phylogenetics and Classification of *Tulipa* (Liliaceae)." *Botanical Journal of the Linnean Society* 172, no. 3 (2013): 280–328.

Curry, Helen Anne. *Evolution Made to Order: Plant Breeding and Technological Innovation in Twentieth-Century America*. Chicago: University of Chicago Press, 2016.

Darwin, Charles. *On the Origin of Species by Means of Natural Selection*. Edited by Harvard Classics. Vol. 11. New York: Collier, 1909.

Dash, Mike. *TulipoMania: The Story of the World's Most Coveted Flower and the Extraordinary Passions It Aroused*. New York: Three Rivers Press, 1999.

Domyan, Eric T., and Michael D. Shapiro. "Pigeonetics Takes Flight: Evolution, Development, and Genetics of Intraspecific Variation." *Developmental Biology* 427, no. 2 (2017): 241–50.

Everson, Ted. *The Gene: A Historical Perspective*. Edited by Brian Baigrie. Greenwood Guides to Great Ideas in Science. Westport, CT: Greenwood Press, 2007.

Hodges, Laurie. *Growing Seedless (Triploid) Watermelons*. Lincoln: Digital Commons@University of Nebraska, 2007. http://extensionpublications.unl.edu/assets/pdf/g1755.pdf.

Howe, Herbert M. "A Root of van Helmont's Tree." *Isis* 56, no. 4 (1965): 408–19.

LeConte, Joseph. *Instructions for the Manufacture of Saltpetre*. Columbia, SC: Charles P. Pelham, State Printer, 1862.

Leigh, G. J. *The World's Greatest Fix: A History of Nitrogen and Agriculture*. Oxford: Oxford University Press, 2004.

Liebig, Justus. *Organic Chemistry in Its Applications to Agriculture and*

Physiology. London: Taylor and Walton, 1840.

McWilliams, James E. *American Pests: The Losing War on Insects from Colonial Times to DDT*. New York: Columbia University Press, 2008.

Murphy, Denis J. *Plants, Biotechnology, and Agriculture*. Oxfordshire, UK: CAB International, 2011.

Olmstead, Alan L., and Paul W. Rhode. "Adapting North American Wheat Production to Climatic Challenges, 1839–2009." *Proceedings of the National Academy of Sciences* 108, no. 2 (2011): 480–85.

Otto, Sarah P. "The Evolutionary Consequences of Polyploidy." *Cell* 131, no. 3 (2007): 452–62.

Pant, Archana P., Theodore J. K. Radovich, Nguyen V. Hue, and Robert E. Paull. "Biochemical Properties of Compost Tea Associated with Compost Quality and Effects on Pak Choi Growth." *Scientia Horticulturae* 148 (2012): 138–46.

Pearson, Calvin, and Amaya Atucha. "Agricultural Experiment Stations and Branch Stations in the United States." *Journal of Natural Resources and Life Sciences Education* 44, no. 1 (2015): 1–5.

Percival, John. *The Wheat Plant: A Monograph*. New York: E. P. Dutton, 1921.

Quisenberry, K. S., and L. P. Reitz. "Turkey Wheat: The Cornerstone of an Empire." *Agricultural History* 48, no. 1 (1974): 98–110.

Rossiter, Margaret W. *The Emergence of Agricultural Science: Justus Liebig and the Americans, 1840–1880*. New Haven, CT: Yale University Press, 1975.

Russell, E. John. "Rothamsted and Its Experiment Station." *Agricultural History* 16, no. 4 (1942): 161–83.

Schlegel, Rolf H. J. *Concise Encyclopedia of Crop Improvement: Institutions, Persons, Theories, Methods, and Histories*. New York: Haworth Press, 2007.

te Beest, Mariska, Johannes J. Le Roux, David M. Richardson, Anne K. Brysting, Jan Suda, Magdalena Kubešová, and Petr Pyšek. "The More the Better? The Role of Polyploidy in Facilitating Plant Invasions." *Annals of Botany* 109, no. 1 (2012): 19–45.

Theunissen, B. "Darwin and His Pigeons: The Analogy between Artificial and Natural Selection Revisited." *Journal of the History of Biology* 45, no. 2 (2012): 179–212.

Van de Peer, Yves, Eshchar Mizrachi, and Kathleen Marchal. "The Evolutionary Significance of Polyploidy." *Nature Reviews Genetics* 18, no. 7 (2017): 411–24.

van der Ploeg, R. R., W. Böhm, and M. B. Kirkham. "On the Origin of the Theory of Mineral Nutrition of Plants and the Law of the Minimum." *Soil Science Society of America Journal* 63, no. 5 (1999): 1055–62.

White, Désirée, and Montserrat Rabago-Smith. "Genotype-Phenotype Associations and Human Eye Color." *Journal of Human Genetics* 56, no. 1 (2011): 5–7.

White, K. D. *Roman Farming*. Edited by H. H. Scullard. Aspects of Greek and Roman Life. Ithaca, NY: Cornell University Press, 1970.

White, Orland E. "Studies in Pisum. I. Inheritance of Cotyledon Color." *American Society of Naturalists* 50, no. 597 (1916): 530–47.

Wood, Andrew R., Dorota Pasko, Michael N. Weedon, Timothy M. Frayling, Zoltán Kutalik, Maija Hassinen, Caroline Hayward, et al. "Defining the Role of Common Variation in the Genomic and Biological Architecture of Adult Human Height." *Nature Genetics* 46, no. 11 (2014): 1173–86.

Wood, David L. "American Indian Farmland and the Great War." *Agricultural History* 55, no. 3 (1981): 249–65.

CHAPTER SEVEN

Arranz-Otaegui, Amaia, Lara Gonzalez Carretero, Monica N. Ramsey, Dorian Q. Fuller, and Tobias Richter. "Archaeobotanical Evidence Reveals the Origins of Bread 14,400 Years Ago in Northeastern Jordan." *Proceedings of the National Academy of Sciences* 115, no. 31 (2018): 7925–30.

Athwal, D. S. "Semidwarf Rice and Wheat in Global Food Needs." *Quarterly Review of Biology* 46, no. 1 (1971): 1–34.

Berkhoff, Karel C. *Harvest of Despair: Life and Death in Ukraine under Nazi Rule*. Cambridge, MA: Belknap Press of Harvard University of Press, 2004.

Bessel, Richard. "The Nazi Capture of Power." *Journal of Contemporary History* 39, no. 2 (2004): 169–88.

Borlaug, Norman E. *Breeding Methods Employed and the Contributions of Norin 10 Derivatives to the Development of the High Yielding Broadly Adapted Mexican Wheat Varieties*. El Batan, Mexico: International Maize and Wheat Improvement Center, 1981.

Borlaug, Norman E. "Sixty-Two Years of Fighting Hunger: Personal Recollections." *Euphytica* 157, no. 3 (2007): 287–97.

Campbell, H. Webster. *Campbell's 1905 Soil Culture Manual: Explains How the Rain Waters Are Stored and Conserved in the Soil*. Lincoln, NE: H. W. Campbell, 1905.

Campbell, H. Webster. *Progressive Agriculture, 1916: Tillage, Not Weather, Controls Yield*. Vol. 155. Lincoln, NE: Woodruff Bank Note Co., 1916.

CIMMYT (International Wheat and Maize Improvement Center). "CIMMYT: Our Work." Home page, accessed August 2, 2019. http://www.cimmyt.org/our-work/.

Clark, J. Allen, John Martin, and Carleton R. Ball. *Classification of American Wheat Varieties*. Edited by the United States Department of Agriculture. Washington, DC: Government Printing Office, 1923.

Collingham, Lizzie. *The Taste of War: World War II and the Battle for Food*. New York: Penguin Press, 2012.

Cronon, William. *Nature's Metropolis: Chicago and the Great West*. New York: W. W. Norton, 1991.

Cullather, Nick. *The Hungry World: America's Cold War Battle against Poverty in Asia*. Cambridge, MA: Harvard University Press, 2010.

Drache, Hiram. "Thomas D. Campbell: The Plower of the Plains." *Agricultural History* 51, no. 1 (1977): 78–91.

Egan, Timothy. *The Worst Hard Time: The Untold Story of Those Who Survived the Great American Dust Bowl*. Boston: Houghton Mifflin Harcourt, 2006.

Finnel, H. H. "The Dust Storms of 1954." *Scientific American* 191, no. 1 (1954): 25–29.

Fleming, Walter C. "Federal Indian Policy: A Summary." In *Visions of an Enduring People: A Reader in Native American Studies*, edited by Walter C. Fleming and John G. Watts, 109–20. Dubuque, IA: Kendall/Hunt, 2009.

Gerhard, Gesine. *Nazi Hunger Politics: A History of Food in the Third Reich*. Lanham, MD: Rowman and Littlefield, 2015.

Hall, Tom G. "Wilson and the Food Crisis: Agricultural Price Control during World War I." *Agricultural History* 47, no. 1 (1973): 25–46.

Hamalainen, Pekka. "The First Phase of Destruction Killing the Southern Plains Buffalo, 1790–1840." *Great Plains Quarterly* 21, no. 2 (2001): 101–14.

Hargreaves, Mary W. M. "Dry Farming Alias Scientific Farming." *Agricultural History* 22, no. 1 (1948): 39–56.

Helfer, Stephan. "Rust Fungi and Global Change." *New Phytologist* 201, no. 3 (2013): 770–80.

Hesser, Leon. *The Man Who Fed the World: Nobel Peace Prize Laureate Norman Borlaug and His Battle to End World Hunger.* Dallas: Durban House, 2006.

Hett, Benjamin Carter. *The Death of Democracy: Hitler's Rise to Power and the Downfall of the Weimar Republic.* New York: Henry Holt, 2018.

Hornbeck, Richard. "The Enduring Impact of the American Dust Bowl: Short- and Long-Run Adjustments to Environmental Catastrophe." *American Economic Review* 102, no. 4 (2012): 1477–1507.

Iijima, Morio, Tomoko Asai, Walter Zegada-Lizarazu, Yasunori Nakajima, and Yukihiro Hamada. "Productivity and Water Source of Intercropped Wheat and Rice in a Direct-Sown Sequential Cropping System: The Effects of No-Tillage and Drought." *Plant Production Science* 8, no. 4 (2005): 368–74.

Jones, Gwyn E., and Chris Garforth. "The History, Development, and Future of Agricultural Extension." In *Improving Agricultural Extension: A Reference Manual*, edited by Burton E. Swanson, Robert P. Bentz, and Andrew J. Sofranko, 3–12. Rome: Food and Agriculture Organization of the United Nations, 1997.

Kirsch, Adam. "The System: Two New Histories Show How the Nazi Concentration Camps Worked." *New Yorker*, April 6, 2015, 77–81.

Kohler, Scott. "The Green Revolution." In *Casebook for The Foundation: A Great American Secret*, edited by Joel L. Fleishman, J. Scott Kohler, and Steve Schindler, 51–57. New York: PublicAffairs Books, 2017.

Lal, R., D. C. Reicosky, and J. D. Hanson. "Evolution of the Plow over 10,000 Years and the Rationale for No-Till Farming." *Soil and Tillage Research* 93, no. 1 (2007): 1–12.

Libecap, Gary D., and Zeynep Kocabiyik Hansen. "'Rain Follows the

Plow' and Dryfarming Doctrine: The Climate Information Problem and Homestead Failure in the Upper Great Plains, 1890–1925." *Journal of Economic History* 62, no. 1 (2002): 86–120.

Lumpkin, Thomas A. "How a Gene from Japan Revolutionized the World of Wheat: CIMMYT's Quest for Combining Genes to Mitigate Threats to Global Food Security." In *Advances in Wheat Genetics: From Genome to Field; Proceedings of the 12th International Wheat Genetics Symposium*, edited by Yasunari Ogihara, Shigeo Takumi, and Hirokazu Handa, 13–20. Tokyo: Springer Japan, 2015.

Mann, Charles C. *The Wizard and the Prophet: Two Remarkable Scientists and Their Dueling Visions to Shape Tomorrow's World*. New York: Alfred A. Knopf, 2018.

Matsusmoto, Takeo. "Norin 10: A Dwarf Winter Wheat Variety." *Japanese Agricultural Research Quarterly* 4 (1968): 22–26.

McDonald, Bryan L. *Food Power: The Rise and Fall of the Postwar American Food System*. New York: Oxford University Press, 2017.

McWilliams, James E. *American Pests: The Losing War on Insects from Colonial Times to DDT*. New York: Columbia University Press, 2008.

Morton, Julius Sterling, Albert Watkins, and George L. Miller. *Illustrated History of Nebraska: A History of Nebraska from the Earliest Explorations of the Trans-Mississippi Region*. Vol. 2. Lincoln, NE: J. North, 1913.

Offer, Avne. *The First World War: An Agrarian Interpretation*. Oxford: Clarendon Press, 1989.

Perkins, John H. *Geopolitics and the Green Revolution*. Oxford: Oxford University Press, 1997.

Reitz, L. P., and S. C. Salmon. "Origin, History, and Use of Norin 10 Wheat." *Crop Science* 8, no. 6 (1968): 686–89.

"Relief Rebus." *Time*, January 9, 1928, 16.

Roser, Max. "Employment in Agriculture," 2019. Accessed August 2, 2019. Published online at OurWorldInData.org. Retrieved from https://ourworldindata.org/employment-in-agriculture.

Russell, Peter A. "The Far-from-Dry Debates: Dry Farming on the Canadian Prairies and the American Great Plains." *Agricultural History* 81, no. 4 (2007): 493–521.

Tauger, Mark B. *Agriculture in World History*. Edited by Peter N. Stearns.

Themes in World History. London: Routledge, 2011.

Widtsoe, John A. *Dry Farming: A System of Agriculture for Countries under Low Rainfall*. New York: Macmillan, 1920.

Wood, David L. "American Indian Farmland and the Great War." *Agricultural History* 55, no. 3 (1981): 249–65.

Worster, Donald. *Dust Bowl: The Southern Plains in the 1930s*. 25th anniversary ed. Oxford: Oxford University Press, 2004.

Wright, Robert J. "Chinch Bug Management." NebGuide, revised November 2013. http://extensionpublications.unl.edu/assets/pdf/g806.pdf.

CHAPTER EIGHT

Alzohairy, Ahmed M., Gábor Gyulai, Robert K. Jansen, and Ahmed Bahieldin. "Transposable Elements Domesticated and Neofunctionalized by Eukaryotic Genomes." *Plasmid* 69, no. 1 (2013): 1–15.

Benbrook, Charles M. "Trends in Glyphosate Herbicide Use in the United States and Globally." *Environmental Sciences Europe* 28 (2016). https://doi.org/10.1186/s12302-016-0070-0.

Bhat, Javaid A., Sajad Ali, Romesh K. Salgotra, Zahoor A. Mir, Sutapa Dutta, Vasudha Jadon, Anshika Tyagi, et al. "Genomic Selection in the Era of Next Generation Sequencing for Complex Traits in Plant Breeding." *Frontiers in Genetics* 7 (2016). https://doi.org/10.3389/fgene.2016.00221.

Cao, Chensi, Jiajia Xu, Guangyong Zheng, and Xin-Guang Zhu. "Evidence for the Role of Transposons in the Recruitment of Cis-regulatory Motifs during the Evolution of C4 Photosynthesis." *BMC Genomics* 17 (2016). https://doi.org/10.1186/s12864-016-2519-3.

Carey, Nessa. *Junk* DNA*: A Journey through the Dark Matter of the Genome*. New York: Columbia University Press, 2015.

Chuong, Edward B., Nels C. Elde, and Cédric Feschotte. "Regulatory Activities of Transposable Elements: From Conflicts to Benefits." *Nature Reviews Genetics* 18, no. 2 (2017): 71–86.

Doolittle, W. Ford, and Carmen Sapienza. "Selfish Genes, the Phenotype Paradigm and Genome Evolution." *Nature* 284, no. 5757 (1980): 601–3.

Douglas, Angela E. "Strategies for Enhanced Crop Resistance to Insect Pests." *Annual Review of Plant Biology* 69, no. 1 (2018): 637–60.

Eaton, Emily. *Growing Resistance: Canadian Farmers and the Politics of Genetically Modified Wheat*. Winnipeg: University of Manitoba Press, 2013.

Evenson, R. E., and D. Gollin. "Assessing the Impact of the Green Revolution, 1960 to 2000." *Science* 300, no. 5620 (2003): 758–62.

Fedoroff, Nina V. "McClintock's Challenge in the 21st Century." *Proceedings of the National Academy of Sciences* 109, no. 50 (2012a): 20200–203.

Fedoroff, Nina V. "Transposable Elements, Epigenetics, and Genome Evolution." *Science* 338, no. 6108 (2012b): 758–67.

Feschotte, Cédric. "Transposable Elements and the Evolution of Regulatory Networks." *Nature Reviews Genetics* 9, no. 5 (2008): 397–405.

Fischer, Iris, Jacques Dainat, Vincent Ranwez, Sylvain Glémin, Jean-François Dufayard, and Nathalie Chantret. "Impact of Recurrent Gene Duplication on Adaptation of Plant Genomes." *BMC Plant Biology* 14, no. 1 (2014). https://doi.org/10.1186/1471-2229-14-151.

Fita, Ana, Adrián Rodríguez-Burruezo, Monica Boscaiu, Jaime Prohens, and Oscar Vicente. "Breeding and Domesticating Crops Adapted to Drought and Salinity: A New Paradigm for Increasing Food Production." *Frontiers in Plant Science* 6 (2015). https://doi.org/10.3389/fpls.2015.00978.

Guo, Ya-Long. "Gene Family Evolution in Green Plants with Emphasis on the Origination and Evolution of *Arabidopsis thaliana* Genes." *Plant Journal* 73, no. 6 (2013): 941–51.

Harre, Nick T., Haozhen Nie, Yiwei Jiang, and Bryan G. Young. "Differential Antioxidant Enzyme Activity in Rapid-Response Glyphosate-Resistant *Ambrosia trifida*." *Pest Management Science* 74, no. 9 (2018): 2125–32.

Heap, Ian. "International Survey of Herbicide Resistant Weeds." Accessed December 12, 2019. http://www.weedscience.org.

Heap, Ian, and Stephen O. Duke. "Overview of Glyphosate-Resistant Weeds Worldwide." *Pest Management Science* 74, no. 5 (2017): 1040–49.

International Wheat Genome Sequencing Consortium. "Shifting the Limits in Wheat Research and Breeding Using a Fully Annotated

Reference Genome." *Science* 361, no. 6403 (2018). https://doi.org/10.1126/science.aar7191.

Kaur, Simerjeet, Kanwarpal S. Dhugga, Robin Beech, and Jaswinder Singh. "Genome-Wide Analysis of the Cellulose Synthase-Like (Csl) Gene Family in Bread Wheat (*Triticum aestivum* L.)." *BMC Plant Biology* 17 (2017). https://doi.org/10.1186/s12870-017-1142-z.

Keller, Evellyn Fox. *A Feeling for the Organism: The Life and Work of Barbara McClintock*. New York: W. H. Freeman, 1983.

Klein, Benjamin, Daniel Wibberg, and Armin Hallmann. "Whole Transcriptome RNA-Seq Analysis Reveals Extensive Cell Type–Specific Compartmentalization in *Volvox carteri*." *BMC Biology* 15 (2017). https://doi.org/10.1186/s12915–017–0450-y.

Klümper, Wilhelm, and Matin Qaim. "A Meta-analysis of the Impacts of Genetically Modified Crops." *PloS One* 9, no. 11 (2014): e111629. https://doi.org/10.1371/journal.pone.0111629.

Kulkarni, Manoj, Raju Soolanayakanahally, Satoshi Ogawa, Yusaku Uga, Michael G. Selvaraj, and Sateesh Kagale. "Drought Response in Wheat: Key Genes and Regulatory Mechanisms Controlling Root System Architecture and Transpiration Efficiency." *Frontiers in Chemistry* 5 (2017). https://doi.org/10.3389/fchem.2017.00106.

Lander, Eric S. "Initial Impact of the Sequencing of the Human Genome." *Nature* 470, no. 7333 (2011): 187–97.

Li, Xinguo, Harry X. Wu, and Simon G. Southerton. "Comparative Genomics Reveals Conservative Evolution of the Xylem Transcriptome in Vascular Plants." *BMC Evolutionary Biology* 10 (2010). https://doi.org/10.1186/1471-2148-10-190.

Liang, Shih Chieh, Ben Hartwig, Pumi Perera, Santiago Mora-García, Erica de Leau, Harry Thornton, Flavia Lima de Alves, et al. "Kicking against the PRCs—a Domesticated Transposase Antagonises Silencing Mediated by Polycomb Group Proteins and Is an Accessory Component of Polycomb Repressive Complex 2." *PLOS Genetics* 11, no. 12 (2015): e1005660. https://doi.org/10.1371/journal.pgen.1005660.

Matt, Gavriel, and James Umen. "Volvox: A Simple Algal Model for Embryogenesis, Morphogenesis and Cellular Differentiation." *Developmental Biology* 419, no. 1 (2016): 99–113.

McClintock, B. "Chromosome Organization and Genic Expression." *Cold Spring Harbor Symposium on Quantitative Biology* 16 (1951): 13–47.

Motta, Erick V. S., Kasie Raymann, and Nancy A. Moran. "Glyphosate Perturbs the Gut Microbiota of Honey Bees." *Proceedings of the National Academy of Sciences* 115, no. 41 (2018): 10305–10.

Nadeau, Nicola J., Carolina Pardo-Diaz, Annabel Whibley, Megan A. Supple, Suzanne V. Saenko, et al. "The Gene Cortex Controls Mimicry and Crypsis in Butterflies and Moths." *Nature* 534, no. 7605 (2016): 106–10.

Orgel, L. E., and F. H. C. Crick. "Selfish DNA: The Ultimate Parasite." *Nature* 284, no. 5757 (1980): 604–7.

Patterson, Eric L., Dean J. Pettinga, Karl Ravet, Paul Neve, and Todd A. Gaines. "Glyphosate Resistance and EPSPS Gene Duplication: Convergent Evolution in Multiple Plant Species." *Journal of Heredity* 109, no. 2 (2018): 117–25.

Pearce, Stephen, Robert Saville, Simon P. Vaughan, Peter M. Chandler, Edward P. Wilhelm, Caroline A. Sparks, Nadia Al-Kaff, et al. "Molecular Characterization of Rht-1 Dwarfing Genes in Hexaploid Wheat." *Plant Physiology* 157, no. 4 (2011): 1820–31.

Pellegrino, Elisa, Stefano Bedini, Marco Nuti, and Laura Ercoli. "Impact of Genetically Engineered Maize on Agronomic, Environmental and Toxicological Traits: A Meta-analysis of 21 Years of Field Data." *Scientific Reports* 8, no. 1 (2018). https://doi.org/10.1038/s41598-018-21284-2.

Perez-Jones, Alejandro, Kee-Woong Park, Nick Polge, Jed Colquhoun, and Carol A. Mallory-Smith. "Investigating the Mechanisms of Glyphosate Resistance in *Lolium multiflorum*." *Planta* 226, no. 2 (2007): 395–404.

Pingali, Prabhu L. "Green Revolution: Impacts, Limits, and the Path Ahead." *Proceedings of the National Academy of Sciences* 109, no. 31 (2012): 12302–8.

Richmond, Todd A., and Chris R. Somerville. "The Cellulose Synthase Superfamily." *Plant Physiology* 124, no. 2 (2000): 495–98.

Sage, Rowan F., Tammy L. Sage, and Ferit Kocacinar. "Photorespiration and the Evolution of C4 Photosynthesis." *Annual Review of Plant Biology* 63 (2012): 19–47.

Sayre, K. D. "Management of Irrigated Wheat." In *Bread Wheat: Improvement and Production*, edited by B. C. Curtis, S. Rajaram, and H. Gómex Macpherson. Rome: Food and Agriculture Organization of the United Nations, 2002. Accessed August 5, 2019. http://www.fao.org/3/y4011e/y4011e00.htm.

Shapiro, James A. "Living Organisms Author Their Read-Write Genomes in Evolution." *Biology* 6, no. 4 (2017). https://doi.org/10.3390/biology6040042.

Shewry, Peter R. "Do Ancient Types of Wheat Have Health Benefits Compared with Modern Bread Wheat?" *Journal of Cereal Science* 79 (2018): 469–76.

Sigman, Meredith J., and R. Keith Slotkin. "The First Rule of Plant Transposable Element Silencing: Location, Location, Location." *Plant Cell* 28, no. 2 (2016): 304–13.

Sinzelle, L., Z. Izsvák, and Z. Ivics. "Molecular Domestication of Transposable Elements: From Detrimental Parasites to Useful Host Genes." *Cellular and Molecular Life Sciences* 66, no. 6 (2009): 1073–93.

Taylor, Austin L. "Bacteriophage-Induced Mutation in *Escherichia coli*." *Proceedings of the National Academy of Sciences* 50, no. 6 (1963): 1043–51.

Tian, Xiuling, Zhanwang Zhu, Li Xie, Dengan Xu, Jihu Li, Chao Fu, Xinmin Chen, et al. "Preliminary Exploration of the Source, Spread, and Distribution of Rht24 Reducing Height in Bread Wheat." *Crop Science* 59, no. 1 (2019): 19–24.

USDA Economic Research Service. "Recent Trends in GE Adoption: Adoption of Genetically Engineered Crops in the U.S." Last modified July 22, 2019. Accessed August 2, 2019. https://www.ers.usda.gov/data-products/adoption-of-genetically-engineered-crops-in-the-us/recent-trends-in-ge-adoption/.

van't Hof, Arjen E., Pascal Campagne, Daniel J. Rigden, Carl J. Yung, Jessica Lingley, Michael A. Quail, Neil Hall, Alistair C. Darby, and Ilik J. Saccheri. "The Industrial Melanism Mutation in British Peppered Moths Is a Transposable Element." *Nature* 534, no. 7605 (2016): 102–5.

Wagoner, Peggy, and Jurgen R. Schaeffer. "Perennial Grain Development: Past Efforts and Potential for the Future." *Critical Reviews in Plant Sciences* 9, no. 5 (1990): 381–408.

Waines, J. Giles, and Bahman Ehdaie. "Domestication and Crop Physiology: Roots of Green-Revolution Wheat." *Annals of Botany* 100, no. 5 (2007): 991–98.

Wickland, Daniel P., and Yoshie Hanzawa. "The Flowering Locus T/Terminal Flower 1 Gene Family: Functional Evolution and Molecular Mechanisms." *Molecular Plant* 8, no. 7 (2015): 983–97.

Workman, Daniel. "Wheat Exports by Country." World's Top Exports, 2019, accessed July 26, 2019. http//www.worldstopexports.com/wheat-exports-country/.

Wulff, B. B. H., and Kanwarpal S. Dhugga. "Wheat—the Cereal Abandoned by GM." *Science* 361, no. 6401 (2018): 451–52.

Zanke, Christine D., Jie Ling, Jörg Plieske, Sonja Kollers, Erhard Ebmeyer, Viktor Korzun, Odile Argillier, et al. "Whole Genome Association Mapping of Plant Height in Winter Wheat (*Triticum aestivum L.*)." *PLoS ONE* 9, no. 11 (2014): e113287. https://doi.org/10.1371/journal.pone.0113287.

CHAPTER NINE

Akagawa, Mitsugu, Tri Handoyo, Takeshi Ishii, Shigenori Kumazawa, Naofumi Morita, and Kyozo Suyama. "Proteomic Analysis of Wheat Flour Allergens." *Journal of Agricultural and Food Chemistry* 55, no. 17 (2007): 6863–70.

Caja, Sergio, Markku Mäki, Katri Kaukinen, and Katri Lindfors. "Antibodies in Celiac Disease: Implications beyond Diagnostics." *Cellular and Molecular Immunology* 8, no. 2 (2011): 103–9.

Catassi, G., E. Lionetti, S. Gatti, and C. Catassi. "The Low FODMAP Diet: Many Question Marks for a Catchy Acronym." *Nutrients* 9, no. 3 (2017). https://doi.org/10.3390/nu9030292.

Cattani, D. J. "Selection of a Perennial Grain for Seed Productivity across Years: Intermediate Wheatgrass as a Test Species." *Canadian Journal of Plant Science* 97, no. 3 (2016): 516–24.

Comino, Isabel, María de Lourdes Moreno, and Carolina Sousa. "Role of Oats in Celiac Disease." *World Journal of Gastroenterology* 21, no. 41 (2015): 11825–31.

Dale, Hanna Fjeldheim, Jessica R. Biesiekierski, and Gülen Arslan Lied. "Non-coeliac Gluten Sensitivity and the Spectrum of Gluten-Related Disorders: An Updated Overview." *Nutrition Research Reviews* 32, no. 1 (2018): 28–37.

De Giorgio, Roberto, Umberto Volta, and Peter R. Gibson. "Sensitivity to Wheat, Gluten and FODMAPs in IBS: Facts or Fiction?" *Gut* 65, no. 1 (2016): 169–78.

DeHaan, Lee, Marty Christians, Jared Crain, and Jesse Poland. "Development and Evolution of an Intermediate Wheatgrass Domestication Program." *Sustainability* 10, no. 5 (2018): 1499–1516.

DeLuca, Thomas H., and Catherine A. Zabinski. "Prairie Ecosystems and the Carbon Problem." *Frontiers in Ecology and the Environment* 9, no. 7 (2011): 407–13.

Federoff, Nina, and Nancy Marie Brown. *Mendel in the Kitchen: A Scientist's View of Genetically Modified Foods*. Washington, DC: John Henry Press, 2004.

Francavilla, R., F. Cristofori, L. Verzillo, A. Gentile, S. Castellaneta, C. Polloni, V. Giorgio, et al. "Randomized Double-Blind Placebo-Controlled Crossover Trial for the Diagnosis of Non-celiac Gluten Sensitivity in Children." *American Journal of Gastroenterology* 113, no 3 (2018): 421–30.

Gibson, Peter R., Gry I. Skodje, and Knut E. A. Lundin. "Non-coeliac Gluten Sensitivity." *Journal of Gastroenterology and Hepatology* 32, no. S1 (2017): S86–S89.

Hayes, C. Richard, Shuwen Wang, T. Matthew Newell, Kathryn Turner, Jamie Larsen, et al. "The Performance of Early-Generation Perennial Winter Cereals at 21 Sites across Four Continents." *Sustainability* 10, no. 4 (2018): 1124–52.

Kantar, Michael B., Catrin E. Tyl, Kevin M. Dorn, Xiaofei Zhang, Jacob M. Jungers, et al. "Perennial Grain and Oilseed Crops." *Annual Review of Plant Biology* 67 (2016): 703–29.

Kasarda, Donald D. "Can an Increase in Celiac Disease Be Attributed to an Increase in the Gluten Content of Wheat as a Consequence of

Wheat Breeding?" *Journal of Agricultural and Food Chemistry* 61, no. 6 (2013): 1155–59.

Larkin, Philip J., Matthew T. Newell, Richard C. Hayes, Jesmin Aktar, Mark R. Norton, Sergio J. Moroni, and Len J. Wade. "Progress in Developing Perennial Wheats for Grain and Grazing." *Crop and Pasture Science* 65, no. 11 (2014): 1147–64.

Li, Hongjie, R. L. Conner, and T. D. Murray. "Resistance to Soil-Borne Diseases of Wheat: Contributions from the Wheatgrasses *Thinopyrum intermedium* and *Th. ponticum*." *Canadian Journal of Plant Science* 88, no. 1 (2008): 195–205.

MacRitchie, F. "Seventy Years of Research into Breadmaking Quality." *Journal of Cereal Science* 70 (2016): 123–31.

Malalgoda, Maneka, and Senay Simsek. "Celiac Disease and Cereal Proteins." *Food Hydrocolloids* 68, no. C (2017): 108–13.

Mohan, Devinder, and Raj Kumar Gupta. "Gluten Characteristics Imparting Bread Quality in Wheats Differing for High Molecular Weight Glutenin Subunits at Glu D1 Locus." *Physiology and Molecular Biology of Plants: An International Journal of Functional Plant Biology* 21, no. 3 (2015): 447–51.

Molina-Infante, Javier, and Antonio Carroccio. "Suspected Nonceliac Gluten Sensitivity Confirmed in Few Patients after Gluten Challenge in Double-Blind, Placebo-Controlled Trials." *Clinical Gastroenterology and Hepatology* 15, no. 3 (2017): 339–48.

Ortolan, Fernanda, and Caroline Joy Steel. "Protein Characteristics That Affect the Quality of Vital Wheat Gluten to Be Used in Baking: A Review." *Comprehensive Reviews in Food Science and Food Safety* 16, no. 3 (2017): 369–81.

Piaskowski, J., Kevin Murphy, Theodore Kisha, and Stephen Jones. "Perennial Wheat Lines Have Highly Admixed Population Structure and Elevated Rates of Outcrossing." *Euphytica* 213, no. 8 (2017). https://doi.org/10.1007/s10681-017-1961-x.

Ribeiro, Miguel, Marta Rodriguez-Quijano, Fernando M. Nunes, Jose Maria Carrillo, Gérard Branlard, and Gilberto Igrejas. "New Insights into Wheat Toxicity: Breeding Did Not Seem to Contribute to a Prevalence of Potential Celiac Disease's Immunostimulatory Epitopes." *Food Chemistry* 213 (2016): 8–18.

Schalk, Kathrin, Barbara Lexhaller, Peter Koehler, and Katharina Anne Scherf. "Isolation and Characterization of Gluten Protein Types from Wheat, Rye, Barley and Oats for Use as Reference Materials." *PloS One* 12, no 2 (2017): e0172819. https://doi.org/10.1371/journal.pone.0172819.

Scherf, Katharina Anne, Peter Koehler, and Herbert Wieser. "Gluten and Wheat Sensitivities—an Overview." *Journal of Cereal Science* 67 (2016): 2–11.

Schulze, Matthias B., Miguel A. Martínez-González, Teresa T. Fung, Alice H. Lichtenstein, and Nita G. Forouhi. "Food Based Dietary Patterns and Chronic Disease Prevention." *BMJ* 361, no. 8157 (2018). https://doi.org/10.1136/bmj.k2396.

Schulz-Schaeffer, J., and S. E. Haller. "Registration of Montana-2 Perennial Xagrothiticum Intermediodurum Khizhnyak." *Crop Science* 27, no. 4 (1987): 822–23.

Shewry, Peter R., and Sandra J. Hey. "The Contribution of Wheat to Human Diet and Health." *Food and Energy Security* 4, no. 3 (2015): 178–202.

Singh, P., A. Arora, T. A. Strand, D. A. Leffler, C. Catassi, P. H. Green, C. P. Kelly, V. Ahuja, and G. K. Makharia. "Global Prevalence of Celiac Disease: Systematic Review and Meta-analysis." *Clinical Gastroenterology and Hepatology* 16, no. 6 (2018): 823–36.

Tyl, Catrin, and Baraem P. Ismail. "Compositional Evaluation of Perennial Wheatgrass (*Thinopyrum intermedium*) Breeding Populations." *International Journal of Food Science and Technology* 54, no. 3 (2019): 660–69.

Van den Broeck, Hetty, Hein C. de Jong, Elma Salentijn, Liesbeth Dekking, Dirk Bosch, Rob Hamer, Ludovicus J. W. J. Gilissen, Ingrid M. van der Meer, and Marinus J. M. Smulders. "Presence of Celiac Disease Epitopes in Modern and Old Hexaploid Wheat Varieties: Wheat Breeding May Have Contributed to Increased Prevalence of Celiac Disease." *Theoretical and Applied Genetics* 121, no. 8 (2010): 1527–39.

Weiner, Jacob. "Applying Plant Ecological Knowledge to Increase Agricultural Sustainability." *Journal of Ecology* 105, no. 4 (2017): 865–70.

World Health Organization. "Malnutrition." Health Topics: Fact Sheets. Accessed August 2, 2019. https://www.who.int/news-room/fact-sheets/detail/malnutrition.

Zhang, Xiaofei, Steven R. Larson, Liangliang Gao, Soon Li Teh, Lee R. DeHaan, Max Fraser, Ahmad Sallam, et al. "Uncovering the Genetic Architecture of Seed Weight and Size in Intermediate Wheatgrass through Linkage and Association Mapping." *Plant Genome* 10, no. 3 (2017). https://doi.org/10.3835/plantgenome2017.03.0022.

Zörb, Christian, Elisabeth Becker, Nikolaus Merkt, Stephanie Kafka, Sarina Schmidt, and Urs Schmidhalter. "Shift of Grain Protein Composition in Bread Wheat under Summer Drought Events." *Journal of Plant Nutrition and Soil Science* 180, no. 1 (2016): 49–55.

Index